Power Systems and Smart Grids

Other related titles:

You may also like

- PBPO167 | Abdelhay A. Sallam and Om P. Malik | Power Grids with Renewable Energy | 2020
- PBPO075 | Biswarup Das | Power Distribution Automation | 2016
- PBPO086 | Federico Milano | Advances in Power System Modelling, Control and Stability Analysis | 2016
- PBPO068 | Juan M. Gers | Distribution System Analysis and Automation | 2013
- PBPO144 | Brahim Aïssa and Nouar Tabet | Photovoltaic Technology for Hot and Arid Environments | 2023

We also publish a wide range of books on the following topics:
Computing and Networks
Control, Robotics and Sensors
Electrical Regulations
Electromagnetics and Radar
Energy Engineering
Healthcare Technologies
History and Management of Technology
IET Codes and Guidance
Materials, Circuits and Devices
Model Forms
Nanomaterials and Nanotechnologies
Optics, Photonics and Lasers
Production, Design and Manufacturing
Security
Telecommunications
Transportation

All books are available in print via https://shop.theiet.org or as eBooks via our Digital Library https://digital-library.theiet.org.

IET ENERGY ENGINEERING 264

Power Systems and Smart Grids

Volume 1: Sizing and optimization of energy systems for communities

Edited by
Ali Mohamed Eltamaly and
Ahmed Abdelhamid Zaki Diab

The Institution of Engineering and Technology

About the IET

This book is published by the Institution of Engineering and Technology (The IET).

We inspire, inform and influence the global engineering community to engineer a better world. As a diverse home across engineering and technology, we share knowledge that helps make better sense of the world, to accelerate innovation and solve the global challenges that matter.

The IET is a not-for-profit organisation. The surplus we make from our books is used to support activities and products for the engineering community and promote the positive role of science, engineering and technology in the world. This includes education resources and outreach, scholarships and awards, events and courses, publications, professional development and mentoring, and advocacy to governments.

To discover more about the IET please visit https://www.theiet.org/

About IET books

The IET publishes books across many engineering and technology disciplines. Our authors and editors offer fresh perspectives from universities and industry. Within our subject areas, we have several book series steered by editorial boards made up of leading subject experts.

We peer review each book at the proposal stage to ensure the quality and relevance of our publications.

Get involved

If you are interested in becoming an author, editor, series advisor, or peer reviewer please visit https://www.theiet.org/publishing/publishing-with-iet-books/ or contact author_support@theiet.org.

Discovering our electronic content

All of our books are available online via the IET's Digital Library. Our Digital Library is the home of technical documents, eBooks, conference publications, real-life case studies and journal articles. To find out more, please visit https://digital-library.theiet.org.

In collaboration with the United Nations and the International Publishers Association, the IET is a Signatory member of the SDG Publishers Compact. The Compact aims to accelerate progress to achieve the Sustainable Development Goals (SDGs) by 2030. Signatories aspire to develop sustainable practices and act as champions of the SDGs during the Decade of Action (2020–30), publishing books and journals that will help inform, develop, and inspire action in that direction.

In line with our sustainable goals, our UK printing partner has FSC accreditation, which is reducing our environmental impact to the planet. We use a print-on-demand model to further reduce our carbon footprint.

British Library Cataloguing in Publication Data

A catalogue record for this product is available from the British Library

ISBN 978-1-83953-986-2 (Volume 1 hardback)
ISBN 978-1-83953-987-9 (Volume 1 PDF)
ISBN 978-1-83953-988-6 (Volume 2 hardback)
ISBN 978-1-83953-989-3 (Volume 2 PDF)
ISBN 978-1-83953-990-9 (2 Volume set hardback)

Typeset in India by MPS Limited
Printed in the UK by CPI Group (UK) Ltd, Eastbourne

Cover image credit: owngarden/Moment via Getty Images

Contents

5 Optimization and sizing of isolated hybrid solar PV/DG/battery energy systems for residential community in hot climate areas of Saudi Arabia 179

Ahmed S. Menesy, Hamdy M. Sultan, Ibrahim O. Habiballah, Mahmoud Kassas and Salah Kamel

6 A novel smart grid concept for a 100% green hybrid energy system 215

Ahmed A. Zaki Diab

About the editors

Ali Mohamed Eltamaly is a distinguished full professor at King Saud University, Saudi Arabia, and Mansoura University, Egypt. He received his PhD degree in Electrical Engineering from Texas A&M University in 2000. His research interests include renewable energy, smart grids, power quality, evolutionary and heuristic optimization techniques, and distributed generation. He has authored or coauthored more than 250 refereed journals, conference papers, books, as well as several patents. He supervised numerous graduate theses and participated in international technical projects. He has received a distinguished professor award, held leadership roles in prestigious journals, and chaired conferences.

Ahmed Abdelhamid Zaki Diab, PhD, is a distinguished associate professor in the Department of Electrical Engineering at Minia University, Egypt. He received his PhD in Electrical Engineering from Novosibirsk State Technical University in 2015. His previous positions include serving as a visiting researcher at Kyushu University, Japan. Among his accolades, he received a State Award for Academic and Scientific Excellence in Engineering Science, Egypt. He has published books and authored or co-authored more than 100 refereed journal articles and conference papers. His research interests include renewable energy, smart grids, AC drives, and heuristic optimization.

Chapter 1

An overview of power systems and smart grid optimization for sustainable cities

Ali M. Eltamaly[1,2], Ahmed A. Zaki Diab[3,4], Amer N. Elghaffar[5], Zeyad A. Almutairi[2,6] and Mohamed A. Ahmed[7]

Abstract

This chapter provides a comprehensive overview of power systems and smart grid optimization techniques for sustainable cities. A detailed comparison between traditional power systems and smart grids is presented, highlighting the key advantages and challenges of smart grid technology. The chapter delves into critical smart grid components, such as energy storage systems, demand-side management, and demand response programs, exploring their roles in enhancing grid flexibility, reliability, and efficiency. The impact of weather and load forecasting on grid operations is also discussed, emphasizing the need for accurate and timely predictions to optimize energy dispatch and system stability. Furthermore, the chapter explores optimal dispatch strategies and optimization algorithms for efficient power flow management in smart grids. A comparative analysis of various optimization techniques, including linear programming, nonlinear programming, mixed-integer programming, and metaheuristic algorithms, is presented. Additionally, a review of available software tools for smart grid design and operation, such as HOMER, RETScreen, and GridLAB-D, is provided. Finally, the chapter addresses essential aspects of smart grid security, including protection systems, cybersecurity measures, and communication protocols, to ensure the reliability and resilience of the grid. The increasing integration of renewable energy sources and electric vehicles introduces new security challenges, necessitating robust cybersecurity measures to protect critical infrastructure and sensitive data. By providing a comprehensive overview of these topics, this study serves as a

[1]Electrical Engineering Department, Mansoura University, Egypt
[2]Sustainable Energy Technology Center, King Saud University, Saudi Arabia
[3]Electrical Engineering Department, Minia University, Egypt
[4]Department of Mechatronics Engineering, Minia National University, Egypt
[5]Alfanar Training Institute, Alfanar Engineering Services, Alfanar Company, Saudi Arabia
[6]Mechanical Engineering Department, King Saud University, Saudi Arabia
[7]Department of Electronic Engineering, Universidad Técnica Federico Santa María, Chile

valuable reference for future research and development in the field of smart grids. Future research directions may include exploring advanced optimization techniques, integrating AI and machine learning, and developing innovative solutions for addressing emerging challenges in smart grid operations.

Keywords: Renewable energy; Smart grids; Demand-side management; Demand response; Energy storage systems; Cybersecurity; Electrical vehicles; Optimal sizing

1.1   Introduction

The relentless pursuit of sustainable energy solutions has brought renewable energy sources (RESs) to the forefront of global discourse. Solar, wind, hydro, and geothermal power, among others, offer a compelling alternative to traditional fossil fuels. By harnessing the power of natural forces, these RESs have the potential to revolutionize the energy landscape, mitigating climate change, stimulating economic growth, and enhancing social well-being. One of the most significant benefits of renewable energy is its ability to reduce greenhouse gas (GHG) emissions. By switching our reliance on fossil fuels, we can significantly decrease the release of carbon dioxide and other harmful pollutants into the atmosphere. This, in turn, helps to mitigate climate change and its associated impacts, such as rising sea levels, extreme weather events, and biodiversity loss. Beyond environmental benefits, RESs can stimulate economic growth and create jobs. Additionally, the development of renewable energy infrastructure can attract investment and drive economic development in both urban and rural areas. Moreover, renewable energy can enhance energy security by reducing dependence on fossil fuel imports. This can lead to more stable energy prices and protect nations from geopolitical risks. Furthermore, the decentralized nature of many RESs, such as rooftop solar panels and small-scale wind turbines, can empower communities and reduce their vulnerability to power outages.

The integration of RESs into the existing power grid presents several technical and economic challenges. One of the primary challenges associated with RESs is their intermittent nature. Solar and wind power generation is heavily dependent on weather conditions, leading to fluctuations in energy output. This variability can disrupt the delicate balance between supply and demand. To address this issue, smart grid technologies have emerged as a critical solution. By leveraging the energy storage systems (ESSs), demand-side management (DSM), and weather and load forecasting (WLF), these challenges can be avoided. Moreover, advanced sensor technologies, coupled with robust communication networks and sophisticated control systems, empower smart grids to monitor and control energy flows in real-time. Another significant challenge is the often remote location of renewable energy projects, particularly wind and solar farms. The transmission of electricity over long distances can result in significant energy losses. Smart grids can mitigate these losses by localizing the power generation at the load locations and optimizing power flow using advanced energy management (EM) strategies.

Smart grids are revolutionizing the way we produce, distribute, and consume electricity. By harnessing advanced technologies, these intelligent power systems offer a multitude of benefits, including enhanced reliability, increased efficiency, and reduced environmental impact. Smart grids also empower consumers by providing real-time energy usage data. This information enables individuals to make informed decisions about their energy consumption, leading to reduced energy bills and a smaller carbon footprint. Additionally, smart grids can facilitate DSM programs, where consumers are incentivized to adjust their energy consumption patterns to be more correlated with the available generation from RESs. Furthermore, smart grids contribute to a cleaner and more sustainable future and help mitigate the impacts of climate change by enabling the transition to a low-carbon economy.

The transition to a smart grid presents a complex array of challenges that must be carefully addressed to realize its full potential. While the benefits of smart grids are substantial, their implementation requires significant technical, economic, and social considerations. One of the primary technical challenges is that the increasing connectivity of smart grid components exposes them to cyber threats. Robust cybersecurity measures must be implemented to protect critical infrastructure and sensitive data [1–5].

From an economic perspective, the initial investment in upgrading the grid infrastructure can be substantial. Governments and utilities must carefully assess the costs and benefits of smart grid projects to ensure a sound return on investment. Moreover, developing a supportive regulatory framework is essential to encourage investment in smart grid technologies and incentivize innovation. Social acceptance is another critical factor in the successful implementation of smart grids. Public education and engagement are necessary to address concerns about privacy, security, and potential job displacement. It is important to emphasize the benefits of smart grids, such as improved reliability, reduced energy costs, and environmental sustainability. Cultural competence and community engagement are very important issues to promote smart grid systems and contribute to a more sustainable future [6].

Smart grids are emerging as a critical infrastructure component for modern cities, enabling efficient, reliable, and sustainable energy distribution. By integrating advanced technologies, smart grids are transforming the way energy is produced, transmitted, and consumed. Traditional power grids are susceptible to disruptions caused by factors such as extreme weather events, equipment failures, and cyberattacks. Smart grids, equipped with advanced monitoring and control systems, can detect and respond to potential issues in real-time, minimizing the impact of disturbances and reducing the frequency and duration of power outages. This increased reliability is essential for critical infrastructure, such as hospitals, transportation systems, and emergency services, which rely on a continuous power supply. Smart grids also empower consumers by providing real-time energy usage data.

The convergence of technological advancements and urban planning has given rise to the concept of smart cities. These cities leverage digital technologies to optimize infrastructure, enhance services, and improve the overall quality of life for their residents. At the heart of these smart cities lies a critical component: smart grids. By integrating digital technologies into the traditional power grid, smart grids enable real-time monitoring, control, and optimization of energy flow. This increased level of intelligence and automation enhances the grid's ability to adapt to changing conditions and meet the evolving needs of modern cities. One of the key benefits of smart grids is their ability to improve grid reliability. By leveraging advanced sensors, data analytics, and automation systems, smart grids can detect and respond to potential issues, such as power outages or voltage fluctuations, in real-time. This proactive approach minimizes disruptions to critical infrastructure and ensures a continuous power supply. Furthermore, smart grids facilitate the integration of RESs, such as solar and wind power. These intermittent sources can be effectively integrated into the grid through advanced technologies like ESSs and DSM programs considering WLF. By optimizing energy flow and balancing supply and demand, smart grids can ensure a stable and reliable power supply, even in the face of fluctuating renewable energy generation. Smart grids also empower consumers by providing real-time energy usage data. This information enables individuals to make informed decisions about their energy consumption, leading to reduced energy bills and a smaller environmental footprint.

Smart grids are composed of several key components such as smart meters, advanced metering infrastructure (AMI), distributed energy resources (DERs) such as solar and wind, advanced communication networks (ACN), control systems, ESS, and loads, as shown in Figure 1.1.

Different concepts in the smart grid used the optimization algorithms in solving different types of problems such as EM [7–9], DSM, increased penetration of

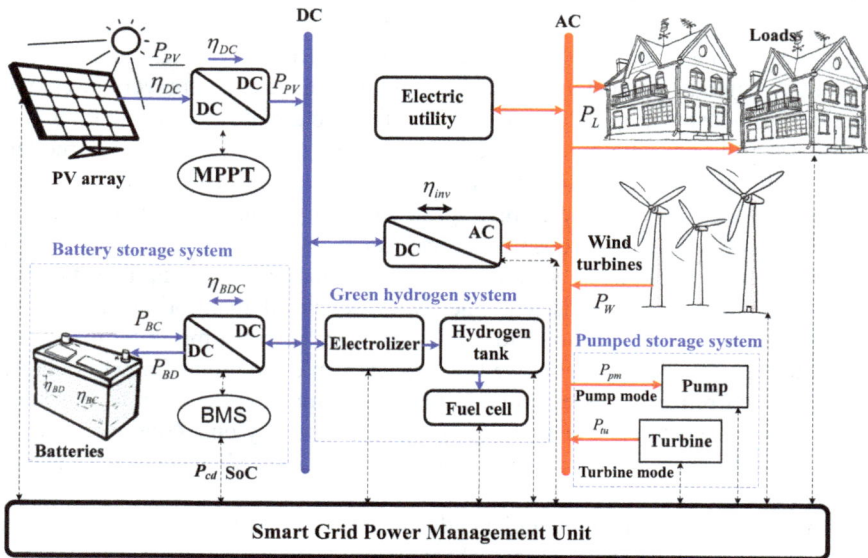

Figure 1.1   The connection of smart grid components

renewable energy sources (IP_RESs), optimal sizing of smart grid components (OS), WLF, fault diagnosis and restoration (FDR), cybersecurity (CyberSec), ACN, and electric vehicle integration (EVI). The flow of power and communication signals between smart grid components is shown in Figure 1.2.

Several review studies have been conducted to provide an overview of smart grid concepts and their application to optimizing various challenges [7,8,10–17]. While these studies offer valuable insights, they typically focus on one or a few specific challenges, such as EM [11,13] or DSM [10,14]. To highlight the unique contribution of this chapter, a comparative analysis with existing review studies is presented in Table 1.1. This comparison reveals that, unlike previous studies, this chapter provides a comprehensive overview of a wider range of smart grid concepts, including operational strategies (OS) and the IP_RESs. It is clear from these review studies that there is no single study covering all the smart grid concepts that have been avoided in this chapter. It is also clear from this table that the EM, DSM, and the MOA are covered in most of the review studies, meanwhile some other concepts such as the OS are covered only in one study [18], and IP_RESs is shown only in [19]. By addressing these under-explored areas, this chapter aims to contribute to a more holistic understanding of smart grid technologies and their potential to enhance the efficiency, reliability, and sustainability of modern power systems.

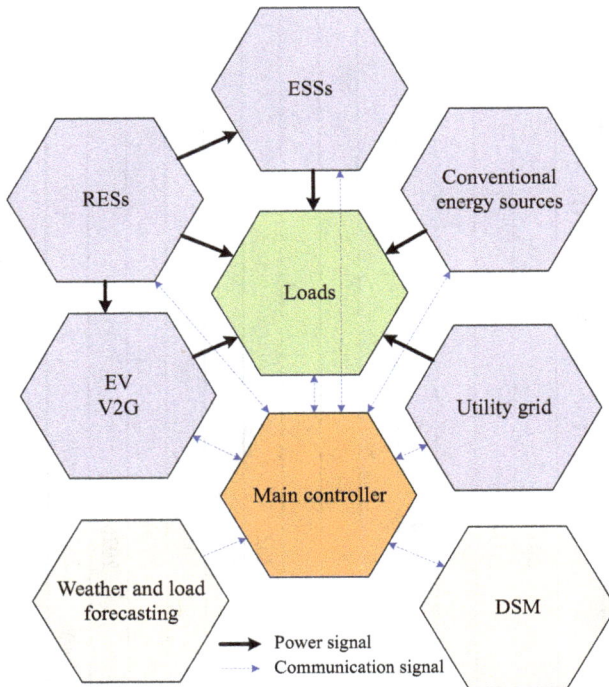

Figure 1.2   The flow of power and communication signals between smart grid components

Table 1.1 The comparison of different review studies in terms of smart grid concepts

Ref.	Year	EM	IP_RESs	OS	WLF	DSM	FDR	CyberSec	ACN	EVI	MOA
[13]	2019										√
[11]	2014					√					
[14]	2018	√				√√					
[19]	2020		√			√					
[15]	2015	√			√	√					
[18]	2017	√		√		√					√
[16]	2022	√√				√					√
[12]	2023								√	√	√
[17]	2024				√√			√√	√√		
[7]	2021	√			√	√			√	√	√
[8]	2020	√			√				√	√	√
[10]	2020	√			√	√		√			√
Proposed study	2025	√√	√√	√√	√√	√√	√√	√√	√√	√√	√√

√√ Deeply focus on the review study.
√ Covered in the review.
X Not covered in the review.

1.2   Energy management

At the heart of the smart grid concepts is EM, which describes the complex process that involves optimizing the load demand and the production from RESs and other power sources. By integrating advanced technologies and intelligent algorithms, EM in smart grids aims to enhance efficiency, reliability, and sustainability. One of the core components of EM in smart grids is the DSM [20–22]. EM systems must be capable of forecasting renewable energy generation and adjusting grid operations accordingly. This involves accurate WLF and real-time monitoring of renewable energy output [7,10,15,17]. ESSs play a crucial role in managing the variability of RESs. By storing excess energy during periods of low demand and releasing it during peak demand periods, ESSs can help balance the smart grid and improve its reliability. Various energy storage technologies, such as batteries, pumped hydro-energy storage (PHES), and compressed air energy storage (CAES) [23], can be integrated into smart grids to enhance their flexibility and resilience [24,25]. A detailed discussion of different types of ESSs is shown in section 1.4.

The EM tools that can be used in the smart grid are shown in Figure 1.3. The smart grid system can use one or more tools to enhance EM. Several studies used ESS to support the stability of the smart grid systems while using flat tariffs without forecasting the weather or loads [26–28]. Some other studies used DSM and ESS for EM without considering the WLF [20,29,30]. Some other studies used all these tools for EM [25,31].

Several studies introduced EM for energy-efficient home energy management controllers based on the DSM and ESSs [15,20–22]. One of these studies introduced a new EM system to optimize home energy consumption [20]. It categorizes appliances and uses a hybrid optimization algorithm (GHSA) to minimize electricity costs and appliance wait times [20]. The system is effective in both single-home and multi-home settings, particularly under different pricing tariffs such as real-time pricing (RTP) and RTEP and critical peak pricing (CPP). Another study [15] introduced a comprehensive review of load forecasting (LF) and dynamic pricing schemes for effective DSM in smart grids. It discussed various LF techniques, including mathematical models like autoregressive, moving average, and exponential smoothing, as well as artificial intelligence (AI)-based models like neural networks and fuzzy logic [15]. Additionally, this study explored different dynamic pricing strategies such as RTP, time-of-use (ToU) pricing, and CPP [15]. Another study [21] introduced a heuristic-based DSM technique to optimize energy consumption in residential areas [21]. The technique aims to minimize electricity costs and peak-to-average ratio. This study compared the performance of different

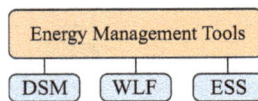

Figure 1.3   EM tools

types of metaheuristic optimization algorithms [21]. Another study [22] proposed a comprehensive framework for EM in smart grids, focusing on optimizing energy production, consumption, and ESS in smart homes. Key contributions of this study include a mixed-integer linear programming (MILP)-based model for smart home energy management and mathematical models for grid-level optimization [22].

1.3   Demand-side management

The DSM is one of the smart grid concepts that have been used to control the loads using different types of activities to make it correlated with the available generation from the RESs of the smart grids. Meanwhile, the demand response (DR) is a DSM mechanism employed to encourage end-use customers to modify their electricity consumption in response to fluctuations in electricity prices or incentives. The other DSM mechanisms are energy efficiency, strategic load growth, and energy competence [6,10–12,14,15,32–35]. The classifications of the DSM based on different mechanisms are shown in Figure 1.4. The following subsections discuss these tools in more detail. The main objectives of the DSM are peak clipping, load shifting, valley filling, load conservations, load building, and flexible load shaping, as shown in Figure 1.5 [29].

Peak power clipping DSM seeks to mitigate peak demand by employing strategies such as ToU pricing to discourage high consumption during peak periods or by implementing direct load control (DLC) to shed non-essential loads [36].

Load shifting DSM aims to flatten the load curve by reducing electricity demand during peak periods and increasing it during off-peak periods. This can be achieved by implementing ToU tariffs, where consumers are charged lower rates for electricity consumed during off-peak hours and higher rates during peak hours [37,38].

Valley filling DSM aims to increase electricity consumption during off-peak periods, often referred to as "valley periods." This can be achieved through various strategies, including offering discounted rates or incentives to consumers to use energy-intensive appliances during these periods. By encouraging increased energy consumption during off-peak hours, utilities can better balance the grid and reduce the need for additional power generation capacity. This strategy can also help to stabilize the grid by preventing voltage fluctuations and improving overall system efficiency [38].

Load conservation DSM focuses on reducing overall energy consumption by promoting energy-efficient practices and technologies. This involves encouraging consumers to adopt energy-saving behaviors, such as turning off lights and appliances when not in use, using energy-efficient appliances, and optimizing heating

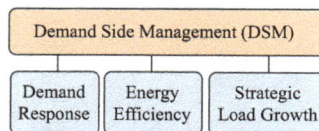

Figure 1.4   The classifications of the DSM based on different mechanisms

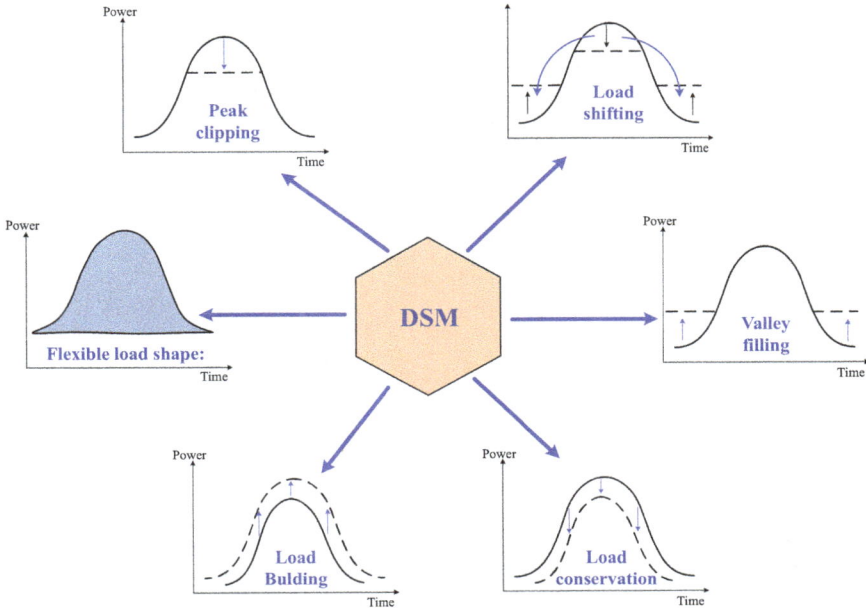

Figure 1.5   DSM strategies

and cooling systems. Utilities can also implement programs to incentivize energy conservation, such as rebates for energy-efficient appliances or ToU pricing. By reducing energy consumption, load conservation DSM helps to alleviate strain on the grid, reduce GHG emissions, and lower energy costs for consumers [39].

Load building DSM involves strategically increasing electricity demand during periods of low demand, often referred to as "valley periods." This can be achieved through various strategies, such as offering discounted rates or incentives to consumers to use energy-intensive appliances during these periods. By encouraging increased energy consumption during off-peak hours, utilities can better balance the grid and reduce the need for additional power generation capacity. This strategy can also help to stabilize the grid by preventing voltage fluctuations and improving overall system efficiency [40].

Flexible load shaping DSM involves optimizing the shape of the load curve by adjusting the timing and magnitude of electricity consumption. This can be achieved through various strategies, such as ToU pricing, RTP, and DR programs. By encouraging consumers to shift their energy consumption from peak to off-peak periods and to reduce their overall energy usage, flexible load shaping can help to balance the grid, improve system reliability, and reduce energy costs [25,30,31,41,42].

1.3.1   Demand response

DR is a mechanism employed to encourage end-use customers to modify their electricity consumption patterns in response to fluctuations in electricity prices or

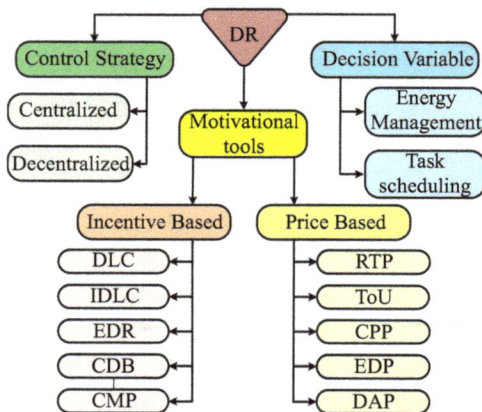

Figure 1.6   The classifications demand response strategies

to give incentive payments designed to induce lower electricity use at times of high market prices [12,43]. The DR can be classified based on motivation tools, decision variables, and control strategy, as shown in Figure 1.6. The following subsections are used to add more details about these classifications.

1.3.1.1   DR classification based on motivation tools

The DR can be classified based on motivation tools into price-based and incentive-based DR strategies, as shown in Figure 1.6 [11]. The price-based DR is subdivided into RTP, ToU pricing, CPP, extreme day pricing (EDP), and day-ahead pricing (DAP). The incentive-based DR can be further classified into incentive-based DR is subdivided into DLC, indirect load control (IDLC), emergency DR (EDR), customer demand bidding (CDB), and central main power (CMP). Detailed review studies were introduced in the literature to compare between all of these DR strategies and detailed in the following subsections [44,45].

The price-based demand response
Price-based DR is a strategy that utilizes time-varying electricity prices to influence consumer behavior. By offering different price levels throughout the day, utilities can incentivize consumers to shift their energy consumption from peak to off-peak hours. This can help to balance the grid, reduce peak demand, and improve overall system efficiency. Price-based DR can be implemented through various pricing mechanisms, such as RTP, ToU pricing, and CPP, among others, as shown in Figure 1.7 [29].

1. **Real-time pricing:** RTP is a DR strategy where the price of electricity fluctuates based on real-time supply and demand conditions. By charging higher prices during peak demand periods and lower prices during off-peak periods, RTP incentivizes consumers to shift their electricity usage to less congested times. This can help reduce peak demand, improve grid stability, and ultimately lower

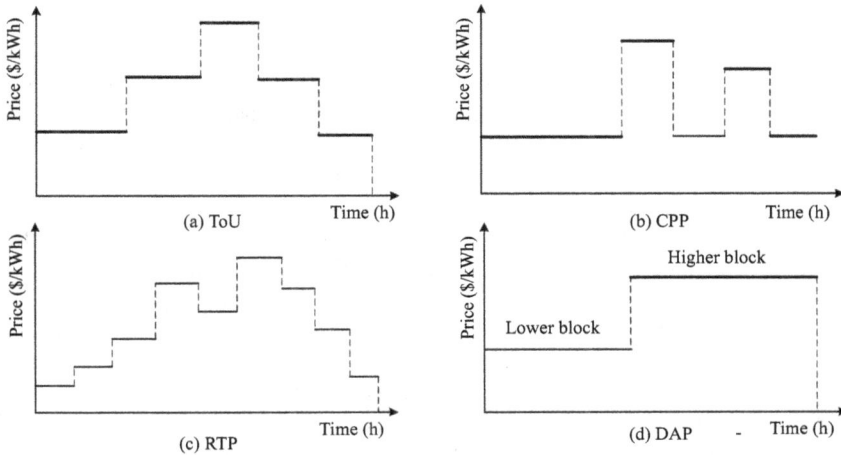

Figure 1.7   The price-based DR programs

overall energy costs. Further enhancement to the RTP can be implemented by considering the current situation of the ESSs and the WLF as introduced in these studies [24,25,30,31,41,42,46]. RTP requires frequent updates to reflect current market conditions. Utilities typically provide RTP information every 15 min to one hour, allowing consumers to make informed decisions about their energy usage. The frequency of updates can vary based on factors like utility infrastructure, consumer preferences, and market conditions.

2. **Time of use (ToU) pricing:** TOU pricing is a DR strategy that divides the day into different pricing periods, typically peak, off-peak, and mid-peak. Consumers are charged different rates for electricity depending on the time of day. By encouraging consumers to shift their energy consumption to off-peak hours, TOU pricing can help reduce peak demand and improve grid efficiency. However, the effectiveness of TOU pricing depends on the flexibility of consumers to adjust their electricity usage. Typically, utilities provide ToU tariffs to customers every month. This allows customers to plan their energy usage in advance and make informed decisions based on the pricing structure. However, some utilities may offer more frequent updates, such as weekly or even daily, especially during periods of high energy demand or significant price fluctuations [47,48].

3. **Critical peak pricing:** CPP is a price-based DR strategy where utilities charge a significantly higher price for electricity during short-duration peak periods. These periods typically occur on hot summer days or during cold winter spells when electricity demand is exceptionally high. By implementing CPP, utilities can encourage consumers to reduce their electricity consumption during these critical times, helping to alleviate stress on the grid and avoid potential blackouts. CPP is typically implemented for short-duration peak periods, often lasting a few hours. Utilities usually provide advance notice of these periods,

often a day or several hours in advance. This allows consumers to adjust their energy usage and reduce their electricity costs during these high-demand times. The specific timing and duration of CPP events can vary depending on factors like weather conditions, seasonal demand, and grid capacity. CPP has been used in different studies to manage the loads during peak periods such as [15,21].

4. **Extreme day pricing:** EDP is a DR strategy where utilities charge significantly higher electricity prices on days with extreme weather conditions, such as extremely hot or cold days. By incentivizing consumers to reduce their energy consumption during these periods, EDP helps to alleviate strain on the grid and prevent blackouts. This pricing mechanism encourages consumers to adopt energy-efficient practices and shift their energy usage to off-peak hours. Utilities usually provide advance notice of these periods, often a day or several days in advance. This allows consumers to adjust their energy usage and reduce their electricity costs during these high-demand times. The specific timing and duration of EDP events can vary depending on weather forecasts and grid conditions. This strategy has been discussed in many studies in the literature [49,50].

5. **Day-ahead pricing:** DAP is a pricing DR mechanism where electricity prices are set for the following day based on forecasted supply and demand conditions. By providing advance notice of price fluctuations, DAP allows consumers to adjust their electricity usage to take advantage of lower prices during off-peak hours. This can help to reduce peak demand and improve grid stability. However, the effectiveness of DAP depends on the accuracy of demand forecasts and the flexibility of consumers to adjust their energy consumption. DAP tariffs are typically provided daily, usually the day before the pricing period begins. This allows consumers to plan their energy usage for the following day based on the forecasted prices. By providing advance notice, DAP empowers consumers to make informed decisions about their energy consumption and potentially reduce their electricity costs [51,52] (Table 1.2).

Incentive-based DR

Incentive-based DR strategies offer direct payments or rewards to consumers for reducing their electricity consumption during peak demand periods or participating in specific demand response events [53]. These incentives can be in the form of cash payments, bill credits, or other valuable rewards. By providing tangible benefits, incentive-based DR programs can motivate consumers to actively manage their energy usage, contributing to a more efficient and reliable grid. Incentive-based DR can be implemented through various pricing mechanisms, such as DLC, IDLC, EDR, CDB, and CMP. These DR strategies are discussed in the following points, and a detailed comparison between these DR strategies is shown in Table 1.3.

1. *Direct load control:* DLC is a DR strategy where utilities directly control specific appliances or devices in consumers' homes to reduce electricity demand during peak periods. This is typically achieved through smart thermostats, programmable switches, or other automated devices that can be remotely controlled by the utility. By remotely turning off or reducing the

Table 1.2   Comparison between price-based DR strategies

Feature	RTP	ToU	CPP	EDP	DAP
Studies	[24,25,30,31]	[47,48]	[15,21]	[49,50]	[51,52]
Frequency of tariff updates	15 min to 1 h	Daily, weekly, or monthly	Short-term, typically hours or days	Short-term, typically days or weeks	Daily
Flexibility	High flexibility for consumers to adjust usage in real-time	Moderate flexibility for consumers to shift usage within pricing periods	Low flexibility, requires immediate response	Low flexibility, requires immediate response	Moderate flexibility for consumers to plan usage based on forecasted prices
Implementation cost	Requires AMI and sophisticated pricing systems	Requires smart meters and billing systems	Relatively low implementation cost, often integrated with existing systems	Relatively low implementation cost, often integrated with existing systems	Requires advanced forecasting and pricing systems
Reliability	Highly reliable, as prices reflect real-time conditions	Reliable, as prices are predictable based on ToU periods	Can be effective in reducing peak demand during critical periods	Can be effective in managing extreme demand conditions	Relies on accurate forecasting and consumer response
Customer satisfaction	Can increase customer engagement and satisfaction, but may lead to price volatility	Can improve customer understanding of energy costs	May lead to temporary inconvenience for consumers	May lead to temporary inconvenience for consumers	Can improve consumer planning and decision-making
Privacy concerns	May require sharing detailed consumption data with the utility	Minimal privacy concerns, as it relies on time-based pricing	Minimal privacy concerns	Minimal privacy concerns	Minimal privacy concerns

power consumption of these devices, utilities can effectively manage peak demand and improve grid stability. While DLC offers significant potential for demand reduction, it requires careful consideration of consumer comfort and privacy concerns. The utility should compensate the customers for switching off their loads in terms of cash payments, bill credits, or other valuable rewards [54].

2. *Indirect load control:* IDLC is a DR strategy that uses incentives and pricing signals to influence consumer behavior rather than directly controlling appliances. By offering incentive-based programs, utilities can encourage consumers to shift their energy consumption to off-peak periods or reduce their overall electricity usage. IDLC provides consumers with more flexibility and

Table 1.3   Comparison between incentive-based DR strategies

Feature	DLC	IDLC	EDR	CDB	CMP
Studies	[54]	[55]	[56]	[57]	[58]
Control mechanism	Utility directly controls specific appliances or devices	Utility indirectly influences consumer behavior through price signals and incentives	Utility triggers emergency demand reduction measures	Consumers actively bid to adjust their energy consumption	Utility incentivizes participants to reduce energy consumption during peak periods
Consumer involvement	Minimal consumer involvement; decisions are made by the utility	Requires active participation from consumers to adjust their energy usage	Requires immediate action from consumers to reduce energy consumption	Requires active participation from consumers to submit bids	Requires participation from organizations to reduce energy consumption
Flexibility	Less flexible, as control is centralized	More flexible, as consumers can adjust their energy usage based on their preferences and needs	Less flexible, as it relies on immediate response from consumers	Highly flexible, as consumers can adjust their bids based on real-time conditions	Less flexible, as it relies on the participation of specific organizations
Privacy concerns	May raise privacy concerns due to direct control of consumer appliances	Generally less intrusive, as it relies on voluntary consumer actions	May require sharing personal information to participate	May require sharing personal information and energy usage data	May require sharing energy usage data with the utility
Reliability	Can provide reliable demand reduction during peak periods	Relies on consumer behavior, which can be less predictable	Can provide rapid demand reduction during emergencies	Relies on consumer participation and bidding behavior	Relies on the participation of specific organizations
Implementation cost	Requires significant investment in smart meters and communication infrastructure	Generally less costly to implement, as it relies on existing infrastructure	Requires effective communication and incentive mechanisms	Requires a robust bidding platform and market infrastructure	Requires effective communication and incentive mechanisms

control over their energy usage while still contributing to grid reliability and efficiency [55].

3. *Emergency DR:* EDR is a critical DR for grid operators to manage sudden and unexpected fluctuations in electricity demand or supply. When grid reliability is threatened, utilities may implement EDR programs to quickly reduce demand by incentivizing consumers to voluntarily curtail their electricity

usage. This can help prevent blackouts, brownouts, and other grid emergencies. EDR programs typically involve short-term, high-value incentives to encourage rapid response from consumers [56].

4. *Customer demand bidding:* CDB DR strategy is consumers actively bid to adjust their electricity consumption in response to RTP signals or incentives. This allows consumers to participate in energy markets and potentially earn rewards for reducing their energy usage during peak demand periods or shifting their consumption to off-peak hours. CDB empowers consumers to take control of their energy usage and contribute to a more efficient and sustainable grid [57].

5. *Central main power:* CMP DR strategy incentivizes participating organizations to reduce their energy consumption during periods of high demand, typically during the summer months. By reducing peak demand, CMP can avoid costly infrastructure upgrades and minimize carbon emissions. Participants in the program receive compensation for their reduced energy usage, promoting a win–win situation for both the utility and its customers [58].

1.3.1.2   DR classification based on decision variable

The DR can be classified based on decision variables into EM and task scheduling, as shown in Figure 1.6. EM DR involves strategically managing electricity demand to optimize grid operations and reduce energy costs. It empowers consumers to adjust their electricity usage in response to price signals or incentives, typically during peak demand periods. By shifting or reducing energy consumption, consumers can help balance the grid, improve system reliability, and lower overall energy costs. Task scheduling DR is a strategy that involves optimizing the timing of non-critical appliances to reduce peak demand and shift load to off-peak periods. By intelligently scheduling appliances like washing machines, dishwashers, and electric vehicle (EV) charging, consumers can contribute to a more balanced and efficient grid. This approach not only benefits individual consumers through potential cost savings but also helps utilities manage overall electricity demand and reduce the need for costly infrastructure upgrades [59].

1.3.1.3   DR classification based on the control strategy

The DR can be classified based on the control strategy into centralized and decentralized DR strategies, as shown in Figure 1.6. The comparison between these two strategies is shown in Table 1.4. The centralized DR involves a central authority, such as a utility or aggregator, making decisions and coordinating the actions of all participants. The central authority collects information about the grid's status, forecasts future demand, and sends control signals to participants to adjust their energy consumption. This approach offers greater control and can lead to significant demand reduction during peak periods. However, it requires a robust communication infrastructure and can be susceptible to single points of failure. Decentralized DR empowers individual consumers to make independent decisions about their energy consumption based on local information and incentives. This approach relies on decentralized communication networks, allowing participants to coordinate and respond to grid conditions without a central authority. Decentralized

Table 1.4   The comparison between the centralized and decentralized DR [60,61]

Feature	Centralized DR	Decentralized DR
Decision-making	Centralized authority makes decisions for all participants	Individual participants make their own decisions based on local information and incentives
Communication	Requires a centralized communication infrastructure to coordinate and control participants	Relies on decentralized communication networks, often peer-to-peer or mesh networks
Complexity	More complex to implement and manage, especially for large-scale systems	Less complex to implement and manage, as decisions are made at the individual level
Scalability	May face scalability challenges as the number of participants increases	More scalable, as it can accommodate a larger number of participants without significant overhead
Privacy	May raise privacy concerns due to the centralized collection and analysis of consumer data	Generally more privacy-friendly, as data is processed locally
Flexibility	Less flexible, as changes to the system require modifications to the central control system	More flexible, as individual participants can adapt their behavior independently
Reliability	More vulnerable to single points of failure in the central control system	More resilient to failures, as decision-making is distributed

DR is more scalable, flexible, and resilient to failures, as decisions are made at the individual level. However, it may require more sophisticated algorithms and communication protocols to ensure efficient coordination and optimization [60,61].

1.3.2   Energy efficiency DSM

Energy efficiency DSM in smart grids focuses on reducing energy consumption through technological advancements and behavioral changes. By implementing energy-efficient appliances, optimizing building designs, and promoting energy-saving practices, consumers and utilities can significantly reduce energy demand. Smart grid technologies, such as AMI and RTP, further enable consumers to monitor and manage their energy usage, leading to reduced energy consumption and lower costs [62].

1.3.3   Strategic load growth DSM

Strategic load growth DSM in smart grids involves managing the growth of electricity demand to ensure the grid's reliability and sustainability. It focuses on identifying and implementing measures to optimize load growth patterns, such as encouraging energy efficiency, promoting DERs, and implementing DR programs. By strategically managing load growth, utilities can avoid costly infrastructure investments, reduce peak demand, and improve overall grid performance [63].

1.3.4 DR mathematical models

Several techniques have been proposed in the literature to model DR programs (DRPs) for controlling load to align with available generation. These DRP mathematical models can be categorized as shown in Figure 1.8 and are described below.

1.3.4.1 Utility function

To model consumer behavior, various utility functions can be employed. A common approach is to use quadratic utility functions, which represent the level of comfort or satisfaction a consumer derives from their energy consumption. This model is further classified into desired load level and marginal benefit models. The desired load level model considers consumers with high power demands, while the marginal benefit model focuses on decreasing satisfaction as energy consumption increases [64].

1.3.4.2 Cost function

The cost function models the cost of generating electricity, often represented as a convex function. This function can be further classified based on how the cost increases with load: either linearly or with increasing marginal costs [65].

1.4 Energy storage systems

The advent of the 21st century has ushered in a new era of technological advancements, transforming industries and reshaping societies. One such transformative technology is the smart grid, a modernized electrical grid system that utilizes digital communication technologies to enhance efficiency, reliability, and sustainability. At the heart of this revolution lies the integration of ESSs, which are poised to revolutionize the way we generate, transmit, and consume energy. Traditional power grids were designed to operate in a unidirectional flow, with power generated at large-scale power plants and distributed to consumers. However, the increasing penetration of RESs, such as solar and wind, has introduced significant challenges due to their intermittent nature. These RESs, while environmentally friendly, can lead to fluctuations in power supply, impacting grid stability and reliability. ESSs offer a promising solution to address these

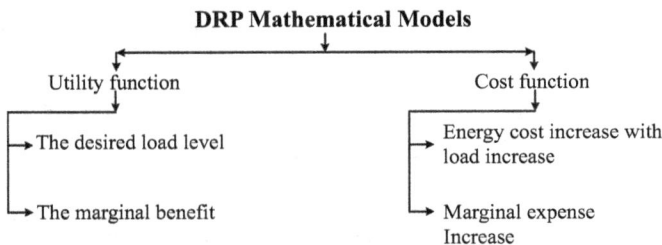

DRP Mathematical Models

Utility function

└──▶ The desired load level

└──▶ The marginal benefit

Cost function

└──▶ Energy cost increase with load increase

└──▶ Marginal expense Increase

Figure 1.8 Classifications of the DRP mathematical models

challenges. By storing excess energy during periods of low demand and releasing it when needed, ESSs can help balance supply and demand, improve grid stability, and integrate RESs more effectively. Furthermore, ESSs can participate in a range of additional benefits, including peak shaving, load shifting, frequency regulation, and voltage control. The integration of ESSs with smart grids has the potential to revolutionize the energy landscape. As technology continues to advance, we can expect to see even greater innovation and deployment of ESSs, further solidifying their role in the evolution of smart grids.

ESSs can be classified based on the function intended to be used as frequency regulation and energy arbitrage. The frequency regulation ESSs are mainly battery, flywheel, and supercapacitor ESSs because of their fast response and their cost-effectiveness in processing a low power at reasonable costs. Meanwhile, energy arbitrage ESSs are mainly used in higher capacity ESSs such as PHES [25,31], CAES [23], and green hydrogen energy systems (GHES) [42]. Both frequency regulation and energy arbitrage are valuable applications of ESSs. By under-standing their specific requirements and selecting appropriate technologies, grid operators can optimize the utilization of energy storage to enhance grid reliability, reduce operational costs, and promote the integration of RESs. Table 1.5 shows a detailed comparison between the frequency regulation and energy arbitrage ESSs in smart grid applications. It is clear from this table that the frequency regulation ESSs are characterized by short response time and low energy capacity.

ESSs can be classified based on the technology used in five categories: electrochemical, mechanical, thermal, electrical storage, and chemical technologies, as shown in Figure 1.9 and detailed in the following points:

Table 1.5   Comparison between the frequency regulation and energy arbitrage ESSs in smart grid applications

Feature	Frequency regulation	Energy arbitrage
Primary function	Maintaining grid frequency stability by rapidly responding to fluctuations in supply and demand	Buying energy at low-price times and selling it at high-price times to profit from price differentials
Response time	Requires rapid response times, often in seconds or milliseconds	Can operate on longer time scales, from minutes to hours
Power rating	High power rating to quickly inject or absorb power into the grid	Moderate to high power rating, depending on the arbitrage strategy
Energy capacity	Low to moderate energy capacity to accommodate frequent, short-duration power injections or absorptions	Higher energy capacity to store energy over longer periods
Technology	Battery, flywheel, and supercapacitors	PHES, CAES, and GHES
Economic considerations	Focuses on providing grid services and receiving payments for frequency regulation	Focuses on maximizing profit by buying energy at low prices and selling at high prices

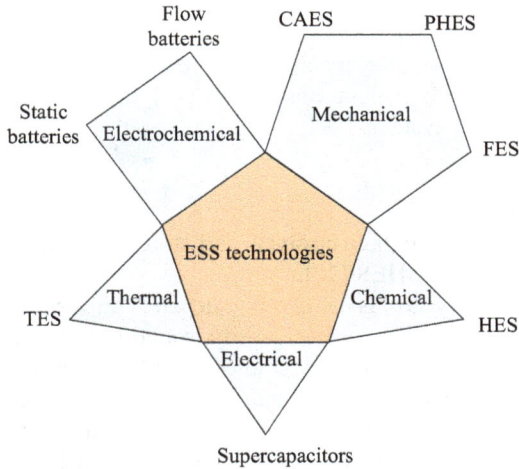

Figure 1.9   Classifications of ESSs based on technologies

1.4.1   Electrochemical storage

○ *Batteries:* This is the most common type of ESSs that store energy in chemical form and convert it into electrical energy when needed. Batteries can be further classified into static and flow batteries, as shown in Figure 1.9. The static batteries include lithium-ion, lead-acid, and flow batteries including vanadium redox flow batteries.

1.4.2   Mechanical storage

This type of ESSs is used to store energy in the mechanical form, and it can be classified into PHES, CAES, and flywheel energy storage (FES) as follows:

○ *Pumped hydro energy storage:* In this type of ESSs, the water is pumped uphill to a reservoir during off-peak hours and released to generate electricity during peak hours. PHES is counted as the cheapest ESSs that can be used in energy arbitrage applications due to its high capacity and long response time [25,29,30].
○ *Compressed air energy storage:* In this type of ESSs, air is compressed and stored in underground caverns or tanks and released to drive turbines during peak demand. CAES can be used in energy arbitrage applications due to its high capacity and low response time [23].
○ *Flywheel energy storage:* Kinetic energy is stored in a rotating flywheel [66].

1.4.3   Thermal energy storage

Heat or cold energy is stored and released as needed, often using materials like molten salt or ice, and it has been used with smart grids in many projects [67,68].

1.4.4   Electrical storage

Electrical storage such as supercapacitors store energy in an electric field and can deliver high power for short durations that is why they are used as a frequency regulation in smart grid systems [69].

1.4.5   Chemical storage

Hydrogen is produced through electrolysis and stored for later use in fuel cells to generate electricity in the GHES [70].

The choice of energy storage technology depends on various factors, including response time, energy capacity, power density, round-trip efficiency, lifespan, and environmental impact. Several review studies have been introduced to discuss the roles of the ESSs in smart grid applications [71–74]. The comparison between different types of ESSs is shown in Table 1.6. It is clear from this table that the supercapacitors, batteries, and FES are suitable for frequency regulation due to their fast response; meanwhile, the rest of ESS technologies are suitable for energy arbitrage applications.

1.4.6   Battery energy storage systems

Batteries are the heart of ESSs used with smart grids due to their fast response, mature technology, and wide range of power and energy capacity [75]. Several technologies used with the smart grid have been reviewed in several studies and compared in Table 1.7. At the heart of these technologies, the lithium-ion batteries (LIBs) have been discussed in many studies [24,76–78]. It is clear from Table 1.7 that LIB and sodium-sulfur (NaS) batteries offer the highest energy and power densities, making them suitable for applications requiring high energy storage and high power output. LIBs generally have the highest efficiency, followed by NaS and flow batteries. Flow and NaS batteries offer significantly longer cycle lives compared to other battery types, making them ideal for long-term, stationary applications. Lead-acid batteries have the lowest cost, but they have lower energy density and shorter cycle life. LIB and flow batteries are more expensive but offer higher performance and longer cycle life. LIBs have a risk of thermal runaway, especially in high-power applications. NaS batteries require high operating temperatures, which can be treated as a good option for hot-environment locations. Flow batteries are generally safer due to their liquid electrolyte. The operating costs are influenced by factors such as charging, maintenance, and replacement. While initial costs may be higher for some technologies, long-term operating costs can be lower due to factors like efficiency and lifespan. Choosing the right battery type depends on the specific application requirements, including energy density, power density, cycle life, safety, cost, and operating environment. As battery technology continues to evolve, we can expect to see further improvements in energy density, power density, cycle life, safety, and cost.

Table 1.6 A detailed comparison between different ESSs considering smart grid applications

Technology	Operating time range	Ramp-up time (s)	Power density (W/kg)	Energy density (Wh/kg)	Efficiency (%)	Life span (Cycles)	Cost per kWh capacity ($) \$	Operating cost ($/kWh)
Supercapacitors	Seconds to minutes	milliseconds	Very high (10,000+)	Low (5–10)	High (90–95)	High (100,000+)	High (500–1000)	Low (0.01–0.05)
Batteries	Minutes to hours	milliseconds	High (100–1000)	Medium to high (100–300)	Medium to high (70–90)	Medium to high (1000–5000)	Medium to high (100–500)	Medium to high (0.05–0.20)
FES	Minutes to hours	<1	Very high (10,000+)	Low (5–10)	High (90–95)	High (100,000+)	High (500–1000)	Low (0.01–0.05)
CAES	Hours to days	60–300	Medium (100–500)	Low (10–20)	Medium (60–70)	High (20,000+)	Medium (200–500)	Medium (0.05–0.15)
PHES	Hours to days	60–600	Low (10–50)	Low to medium (10–50)	High (70–80)	High (30,000+)	Low (50–100)	Low (0.01–0.05)
Green hydrogen energy storage	Hours to days, even weeks	60–300	Medium (100–500)	Low (10–20)	Medium to low (60–70)	High (20,000+)	High (300–1000)	Medium to high (0.10–0.30)
Thermal energy storage	Hours to days	60–3600	Low (10–50)	Low to medium (10–50)	Medium to high (70–80)	High (30,000+)	Medium (100–500)	Medium (0.05–0.15)

Table 1.7 The comparison between different types of batteries used with smart grids

Battery type	Energy capacity (Wh/kg)	Power density (W/kg)	Efficiency (%)	Life span (cycles)	Cost per kWh capacity ($) \$	Safety	Operating cost (/kWh)
Lead-acid	Low (30–50)	Low	Medium (70–80)	Low (300–500)	Low (50–100)	High	Medium
Nickel-cadmium (NiCd)	Medium (45–80)	Medium	Medium (70–80)	Medium (1000–2000)	Medium (100–200)	Medium	Medium
Nickel-metal hydride (NiMH)	Medium–high (60–120)	Medium–high	Medium (70–80)	Medium–high (500–1000)	Medium–high (150–250)	Medium	Medium
Lithium-ion (Li-ion)	High (150–190)	High	High (90–95)	Medium–high (500–1000)	Medium–high (150–300)	Medium (fire risk)	Medium–high
Flow batteries	Medium–high	Medium–high	Medium–high (70–80)	High (10,000+)	High (200–500)	High	Medium
Sodium-sulfur (NaS)	High (150–200)	High	High (80–90)	High (20,000+)	High (200–500)	Low (high-temperature operation)	Medium

1.5   Optimal sizing of smart grid components

A critical aspect of smart grid design is the optimal sizing of its components. This involves determining the appropriate capacity and configuration of various elements, such as generation sources, ESSs, and distribution infrastructure. AI techniques can accurately predict future energy demand and renewable energy generation, leading to more precise sizing decisions. By considering uncertainties and variability, AI can help design more reliable smart grid systems. Optimized sizing can lead to lower capital and operational costs. AI-powered smart grid can adapt to changing conditions, such as variations in energy demand and renewable resource availability. Optimal sizing is a complex task that requires careful consideration of numerous factors shown in the following subsections.

1.5.1   Load demand data

Understanding the diverse and dynamic nature of electricity demand, including residential, commercial, and industrial loads, is crucial for optimal smart grid design. Accurate LF, which considers future load growth and changes in load patterns, is essential for sizing grid components appropriately. If detailed hourly or sub-hourly load data is available, time-series analysis techniques [25,27,29] can be employed to predict future load profiles with high precision. This enables a more accurate assessment of the grid's performance and the sizing of components. However, if such detailed data is unavailable, probabilistic methods like Monte Carlo simulation (MCS) or stochastic programming [79–83] can be used to analyze uncertainties in load patterns and generate various scenarios for robust design. These methods can help in making informed decisions about the sizing of components, even in the absence of precise load data.

1.5.2   Weather data

Weather data such as wind speed, solar irradiances, temperature, etc., is similar to the load demand data and can characterize the type of sizing technique used. Where, in case the hourly or sub-hourly data is available, the time series analysis can be used in the operation of the system in power dispatch such as in [25,27,29,84–87]. Otherwise, MCS or stochastic programming can be used [79–83].

1.5.3   Energy storage systems

Determining the optimal capacity and type of ESSs to balance supply and demand, improve grid reliability, and support RES integration is the main challenge of smart grid sizing. An accurate model for the ESSs considering a detailed degradation [24,25], *SoH*, and economical factors associated with the ESSs is crucial in the optimal sizing of smart grids [25–28]. Moreover, the dispatch strategy used to manage the power flow between the generation from different power sources, loads, and ESSs is very important for accurate results [27,29].

1.5.4   Incorporating DSM

This section delves into the importance of incorporating DSM in the sizing of smart grid components, particularly ESS. Overlooking DSM in the sizing of smart grid components can lead to suboptimal designs and increased costs [25]. By considering DSM strategies, it is possible to reduce peak demand, improve load factor, and optimize the overall system design. By accurately estimating the price elasticity of demand (*PED*) and conducting sensitivity analyses, utilities can assess the impact of DSM on the sizing of grid components. *PED* measures the responsiveness of electricity demand to changes in price. It is defined as the percentage change in demand divided by the percentage change in price, as shown in the following equation:

$$PED = \frac{\% \text{ Change on demand}}{\% \text{ Change on tariff (Price)}} \tag{1.1}$$

PED is a measure of how sensitive the quantity demanded of a good or service is to changes in its price. Typically, *PED* is negative, indicating an inverse relationship between price and demand where, as price increases, demand decreases, and vice versa. This principle aligns with the concept of DR, where consumers adjust their electricity consumption in response to price signals. When tariffs rise, consumers may reduce their load, and conversely, they may increase consumption when prices fall. The magnitude of *PED* determines the elasticity of demand. Inelastic demand (*PED* between 0 and -1) signifies a relatively low responsiveness to price changes, while elastic demand (*PED* between -2 and -1) indicates a higher sensitivity to price. When *PED* equals -1, the demand is unified elastic, meaning a change in price leads to an equal percentage change in quantity demanded, as shown in Figure 1.10 [30].

Several factors influence the *PED* such as

- *Availability of substitutes:* If there are many close substitutes available, demand is likely to be more elastic, as consumers can easily switch to alternative products.
- *Necessity vs. luxury:* Necessary goods, such as food and medicine, tend to have inelastic demand, as consumers are less likely to reduce their consumption

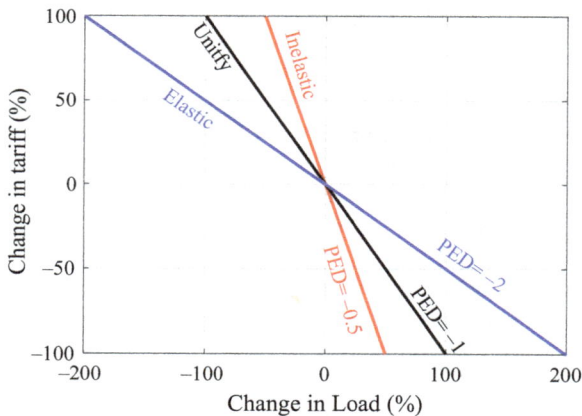

Figure 1.10   Types of PED based on the value

even if prices increase. Luxury goods, on the other hand, tend to have elastic demand.

• *Time horizon:* In the short run, demand may be less elastic as consumers may not have time to adjust their consumption patterns. However, in the long run, demand may become more elastic as consumers have more time to find substitutes or adjust their behaviour.

• *Proportion of income spent:* Goods that represent a small proportion of a consumer's income tend to have inelastic demand, as price changes have a minimal impact on overall spending.

1.5.5   Sensitivity analysis

Sensitivity analysis is a powerful technique used to assess the impact of uncertainties on the design and operation of complex systems, such as smart grids. By systematically varying input parameters, sensitivity analysis helps identify critical factors and potential risks associated with different design choices. Several other uncertainties can impact the design and operation of smart grids such as *PED* uncertainty, WLF uncertainty, renewable energy generation uncertainty, grid components price uncertainty, and policy and regulatory changes.

Several techniques can be used to perform sensitivity analysis in sizing smart grids as follows.

1.5.5.1   One-factor-at-a-time analysis

OAT analysis is a straightforward technique used to assess the sensitivity of a model's output to changes in its input parameters. In this method, one parameter is varied while all other parameters are held constant. By observing the resulting changes in the output, analysts can gain insights into the relative importance of each parameter. Detailed steps of the OAT are shown in the following points:

• *Identify key parameters:* The first step is to identify the key parameters that may significantly impact the model's output. These parameters could be physical, economic, or environmental factors.

• *Base case analysis:* A baseline simulation is conducted using a set of baseline parameter values.

• *Parameter variation:* Each key parameter is varied individually while keeping other parameters fixed. The range of variation can be determined based on expert judgment or historical data.

• *Model simulation:* For each parameter variation, the model is re-run to obtain a new output.

• *Comparison and analysis:* The results from the base case and the varied cases are compared to assess the sensitivity of the output to changes in the parameter.

While OAT analysis is a simple and intuitive technique, it has some limitations such as

• *Interaction effects:* OAT analysis may not capture interactions between parameters. For example, the impact of one parameter may depend on the value of another parameter.

- *Computational efficiency:* For models with many parameters, OAT analysis can be computationally expensive, as each parameter needs to be varied individually.
- *Limited insights:* OAT analysis may not provide a comprehensive under-standing of the uncertainty in the model's output, especially when multiple parameters are uncertain.

Despite its limitations, OAT analysis can be a useful tool for initial sensitivity analysis, especially when the number of parameters is relatively small. This sensitivity analysis has been introduced in many literature of sizing smart grids and hybrid renewable energy systems (HRES) [24,25,27,28,42,76,88–92]. The price variation of energy system components is varied to perform OAT analysis in this study [88]. A detailed OAT analysis is introduced in [89] for integrating renewable energy systems with IEEE 30-bus systems.

1.5.5.2   MCS for sensitivity analysis

MCS is a versatile statistical technique that can be applied to a wide range of problems, including uncertainty analysis in smart grid design. By simulating random outcomes, MCS provides a probabilistic approach to understanding the impact of uncertainty on system performance. By leveraging MCS, engineers and planners can make more informed decisions about the design and operation of smart grids, leading to more resilient, efficient, and sustainable systems. The steps of using MCS for sensitivity analysis are shown in the following points:

- *Identify uncertain parameters:* The first step is to identify the key uncertain parameters that may affect the system's behavior. These parameters could include load demand, renewable energy generation, component failure rates, and economic factors like energy prices.
- *Define probability distributions:* For each uncertain parameter, a probability distribution is assigned to represent the range of possible values and their likelihood. Common probability distributions used in MCS include:
- *Normal distribution:* Used to model continuous variables with a bell-shaped curve.
- *Uniform distribution:* Used to model variables with equal probability over a specified range.
- *Log-normal distribution:* Used to model variables with a skewed distribution, such as wind speed or solar irradiance.
- *Triangular distribution:* Used to model variables with a known minimum, maximum, and most likely value.
- *Generate random samples:* Random values are sampled from the specified probability distributions for each uncertain parameter.
- *Run simulations:* The model is run multiple times, each time using a different set of randomly sampled parameter values.
- *Analyze results:* The results from the multiple simulations are analyzed to determine the range of possible outcomes and the probability of different events. Key performance indicators such as system reliability, operational costs, and environmental impact can be evaluated.

The use of MCS in the sensitivity analysis of sizing smart grids is showing effective results that should be considered such as

- *Comprehensive uncertainty analysis:* MCS can capture the complex interactions between multiple uncertain parameters, providing a more comprehensive understanding of the system's behavior.
- *Flexibility:* It can be applied to a wide range of models and systems, including those with nonlinear and stochastic components.
- *Visualizations:* The results of MCS can be visualized using histograms, scatter plots, and other graphical techniques to help identify trends and patterns.
- *Risk assessment:* By analyzing the probability distribution of key performance indicators, MCS can help identify potential risks and opportunities.

MCS has been applied for sensitivity analysis in smart grid design [93–95]. MCS can be used to assess the impact of uncertainties in load demand, renewable energy generation, and energy prices on the optimal size and operation of ESSs. Moreover, by simulating a large number of scenarios, MCS can be used to estimate the reliability of the power system and identify potential vulnerabilities. Furthermore, MCS can be used to analyze the economic viability of different investment options, considering uncertainties in future energy prices, technological advancements, and policy changes.

1.5.5.3 Latin hypercube sampling (LHS) for sensitivity analysis

LHS is a statistical technique used to improve the efficiency of MCS. By ensuring that all combinations of parameter values are sampled, LHS can provide more accurate results with fewer simulations. By leveraging LHS, engineers and planners can make more informed decisions about the design and operation of smart grids, leading to more resilient, efficient, and sustainable systems. LHS offers several advantages over simple random sampling. By stratifying the parameter space, LHS ensures that all regions of the parameter space are adequately sampled, resulting in more efficient and precise estimates. This reduced variance in the output leads to more reliable conclusions. Additionally, LHS is a flexible technique that can be applied to a wide range of models and systems, including those with complex, nonlinear, and stochastic components.

LHS can be applied to various aspects of smart grid design [96,97]. It can be used to quantify uncertainty in model predictions, such as load forecasts or renewable energy generation forecasts. By identifying critical parameters and their potential impact on system performance, LHS helps assess the risks associated with different design choices. Additionally, LHS can be employed to optimize the design of smart grid components, such as ESSs, by exploring a wide range of parameter combinations.

The steps of LHS application in sensitivity analysis of smart grid sizing and operation are shown in the following points:

- *Partitioning the parameter space:* The range of each uncertain parameter is divided into a specified number of equally probable intervals.

- *Random sampling:* A random value is selected from each interval for each parameter.
- *Stratified sampling:* The selected values are then paired in a stratified manner to ensure that all combinations of parameter values are represented.

1.5.6   Weather and load forecasting

WLF is a critical component of modern smart grid systems. By accurately predicting future energy demand and supply, utilities can make informed decisions about resource allocation, grid operations, and energy market participation. This, in turn, leads to a more reliable, efficient, and sustainable energy system. The smart grid should consider the WLF to change the tariff and control the smart grid to face any abnormal weather or load consequences in the coming hours during the sizing study [7,10,15,17,25,31]. When sizing smart grid components, it is essential to consider the impact of weather and load variability. By incorporating WLF into the sizing process, utilities can get the following benefits:

- *Optimize component sizing:* Accurate forecasts of peak demand and renewable energy generation can help determine the optimal size of components, avoiding overinvestment and underinvestment. For instance, by predicting periods of high renewable energy generation, utilities can size ESSs to store excess energy for later use, reducing the need for additional generation capacity.
- *Improve grid reliability:* By anticipating extreme weather events, such as heat waves, cold spells, and storms, utilities can take proactive measures to maintain grid stability and prevent outages. For example, by forecasting high demand during a heat wave, utilities can implement demand response programs to reduce peak load and avoid potential blackouts.
- *Enhance grid flexibility:* WLF can help identify opportunities to optimize the operation of flexible resources, such as demand response programs and energy storage, to balance supply and demand. By predicting periods of high or low demand, utilities can incentivize consumers to shift their consumption patterns or utilize energy storage to smooth out load fluctuations.
- *Facilitate grid integration of renewable energy:* Accurate forecasts of renewable energy generation can help utilities integrate RESs into the grid more efficiently and reliably. By predicting the output of wind and solar power plants, utilities can adjust their operations to accommodate variable renewable energy generation.

While WLF has made significant progress in recent years, several challenges remain as shown in the following points:

- *Uncertainty in weather forecasts:* Weather forecasts are inherently uncertain, especially for long-term predictions. Factors such as climate change and extreme weather events can further complicate accurate forecasting.
- *Variability in load patterns:* Load patterns can be influenced by various factors, including economic conditions, social behavior, and technological advancements. For example, the increasing adoption of EVs can significantly impact electricity demand, especially during peak hours.

- *Impact of climate change:* Climate change can lead to more frequent and intense extreme weather events, such as heat waves, droughts, and floods. These events can have a significant impact on energy demand and supply, making accurate long-term forecasting even more challenging.

To address these challenges, advanced forecasting techniques, such as machine learning (ML) and AI, are being employed to improve the accuracy and reliability of WLF. By incorporating these techniques, utilities can make more informed decisions and optimize the performance of their smart grids. Many studies incorporated forecasting factor (*FF*) in determining the optimal RTP demand response strategy [7,10,15,17,25,31], as shown in the following equation. The value of this factor gives the percentage of the difference between the generation and the load for the next hours (N_F) in case of N_F equals 24 h so the forecasting is one day ahead. The nearest hour takes a higher weight than the farthest hours by dividing the percentage difference by the counter i. The higher the value of *FF*, the more stable the system in the near future, which means the tariff can be reduced and vice versa.

$$FF(t) = \sum_{i=1}^{N_F} \frac{P_G(t+i) - P_{Lo}(t+i)}{P_{LA} \cdot i} \tag{1.2}$$

where $P_G(t+i)$ and $P_{Lo}(t+i)$ are the generated power at hour $t+i$, respectively. P_{LA} is the average load power.

1.5.7   Objective function

The objective function is a mathematical expression that quantifies the performance of the smart grid. The choice of objective function significantly influences the optimal sizing and operation of smart grid components. A well-defined objective function guides the optimization process toward a solution that best aligns with the specific goals and constraints of the system. There are common objective functions in smart grid design that can balance the design between economic, environmental, and technical factors. Each one of these factors is called a single objective function; meanwhile, the summation of more than one of these single objective functions with different weight values is called a multi-objective function which should be optimized using a suitable optimization algorithm. The detailed description of these single objective functions is discussed in the following subsections.

1.5.7.1   Economic objectives in smart grid design

Economic considerations are paramount in the design and operation of smart grids. The primary economic objectives are to minimize costs and maximize revenue.

Minimizing total cost of ownership (TCO) [98], which represents the total cost of the smart grid over its entire lifecycle. The TCO includes consists of initial investment costs, operational costs, and maintenance costs. By minimizing TCO, utilities can reduce their overall expenditure and improve their financial performance.

Minimizing the levalized cost of energy (LCOE) by dividing the net present cost (NPC) by the yearly load demand energy multiplied by the capital recovery factor (*CRF*) as shown in the following equation [87,99–101]:

$$LCOE = \frac{NPC * CRF}{YE} \tag{1.3}$$

where *YE* is the yearly demand energy consumed, and *CRF* is the capital recovery factor, which can be determined from the following equation:

$$CRF = \frac{r(1 + r)^T}{(1 + r)^T - 1} \tag{1.4}$$

where *T* is the project lifetime in years, and *r* is the net interest rate.

Maximizing revenue through various strategies such as

- *Energy sales:* By efficiently managing energy resources and optimizing pricing strategies, utilities can increase revenue from energy sales.
- *Ancillary services:* Utilities can provide ancillary services, such as voltage control, frequency regulation, and reactive power support, to grid operators.
- *Demand response programs:* By implementing demand response programs, utilities can reduce peak demand and shift load to off-peak periods, leading to cost savings and increased revenue.
- *Energy efficiency programs:* By promoting energy efficiency, utilities can help reduce energy consumption and lower overall demand, leading to reduced operating costs and increased revenue.

1.5.7.2   Environmental objectives in smart grid design

By embracing sustainable practices and technologies, smart grids can significantly contribute to a cleaner and healthier planet. This environmental imperative should be integrated into the design phase of smart grids. For existing fossil fuel power plants, OS can be optimized to minimize their emissions. This can be achieved through techniques like load shifting, curtailment, and coordinated operation with RESs. Furthermore, incorporating environmental considerations into the objective function of smart grid design can help ensure that sustainability is prioritized. This involves balancing economic, technical, and environmental factors to find optimal solutions that minimize carbon emissions, improve air quality, and reduce reliance on fossil fuels. By prioritizing environmental objectives, smart grids can play a crucial role in mitigating climate change and promoting a sustainable future. The primary environmental objective of smart grids is to reduce GHG emissions. This can be achieved through:

- *Promoting renewable energy integration:* Smart grids can facilitate the integration of RESs, such as solar and wind power, by providing advanced control and monitoring systems.
- *Improving energy efficiency:* By implementing energy efficiency measures, such as smart metering and demand response programs, smart grids can reduce energy consumption and associated emissions.

- *Optimizing energy storage:* ESSs can be used to store excess renewable energy for later use, reducing reliance on fossil fuel-based power plants.
- *Minimizing carbon emissions:* This objective seeks to reduce GHG emissions by promoting RESs and energy efficiency measures.

1.5.7.3   Technical objectives in smart grid design

Technical objectives are very important issues that take care of improving the operating performance of the smart grid. These objectives are maximizing the system reliability and power quality such as maintaining voltage and frequency within acceptable limits.

1.5.7.4   Multi-objective optimization in smart grid design

Often, multiple objectives need to be considered simultaneously. This leads to multi-objective optimization problems, where the goal is to find a solution that balances conflicting objectives. Multi-objective optimization involves finding a solution that balances multiple objectives. Techniques such as Pareto optimization and weighted sum methods can be used to address multi-objective problems. A key concept in multi-objective optimization is Pareto optimality. A solution is considered Pareto optimal if it is not possible to improve one objective without worsening another. In other words, a Pareto optimal solution represents a trade-off between different objectives.

Several techniques can be used to address multi-objective optimization problems in smart grid design. One of these techniques is the weighted sum method involves assigning weights to each objective, creating a single objective function. The weights determine the relative importance of each objective. By varying the weights, different trade-offs between objectives can be explored, allowing for a flexible approach to decision-making.

1.5.7.5   The role of weights in objective functions

The relative weights assigned to each objective shape the final solution. For example, if a higher weight is assigned to reliability, the optimization process will prioritize solutions that enhance system reliability, even if it means higher costs. Conversely, if cost is the primary concern, the optimization process may favor solutions that minimize initial investment costs, even if it compromises reliability.

By carefully selecting and weighing the objectives, engineers and planners can design smart grids that are not only cost-effective but also reliable, sustainable, and resilient.

1.5.8   Considering vehicle-to-grid application in smart grid sizing

The integration of vehicle-to-grid (V2G) technology into smart grid systems offers significant potential for enhancing grid flexibility, reliability, and efficiency. V2G technology enables EVs to act as both consumers and suppliers of electricity, allowing them to discharge energy back into the grid during periods of high demand or low renewable energy generation. When sizing smart grid components, such as ESSs and transmission lines, V2G can be considered a flexible resource to reduce the required capacity. However, incorporating V2G into the design process can increase the complexity of the optimization problem. To address the challenges and maximize the benefits of V2G, advanced optimization techniques, such as MILP and dynamic programming, can be employed. These techniques can help determine the optimal sizing and operation of V2G-enabled smart grids, considering factors such as battery degradation, charging/discharging rates, and grid constraints [77,102].

1.5.9   Types of analysis used for smart grid sizing

Considering all the above factors and challenges in the modeling of the smart grid can make success sizing for smart grid components. Any missing information and data in the design process can cause catastrophic situations in terms of reliability and economic outcomes. Due to the wide spread factors that can affect the sizing of smart grid components, it needs an effective optimization algorithm to accurately determine the size of the components within the shortest time. Several optimization techniques are employed to determine the optimal sizing of smart grid components are introduced and discussed in section 1.9. Moreover, considering the numerous factors and challenges involved in modeling smart grids, accurate sizing of components is crucial for ensuring reliable and cost-effective operation. Incomplete or inaccurate data can lead to suboptimal designs and potential system failures. To address these complexities, effective optimization algorithms are essential to determine the optimal size of components within a reasonable timeframe. Several techniques are employed for this purpose, including the following.

1.5.9.1   Analytical techniques

While analytical methods, as employed in References [103–106], can provide valuable insights into the reliability of hybrid power systems, their application to complex systems with numerous random parameters can be computationally demanding and may not accurately capture the stochastic nature of component failures. As a result, alternative approaches, such as simulation-based methods, are becoming increasingly popular for assessing the reliability of these systems [107].

1.5.9.2   Iterative techniques

Iterative techniques are commonly employed for sizing HRES. These techniques involve a trial-and-error approach, where initial estimates for component sizes are made, and then simulations are performed to assess the system's performance. Based on the simulation results, the component sizes are adjusted, and the process is repeated until an optimal solution is found. Several studies have utilized iterative techniques, including [108–112], to effectively size HRES components.

1.5.9.3   Probabilistic techniques

Probabilistic techniques are essential for accurately sizing smart grids due to the inherent uncertainty in RESs like solar and wind. These techniques account for variability in resource availability, load demand, and component performance, leading to more reliable and cost-effective system designs. Here are some of the most common probabilistic techniques used in HRES sizing:

1. *Monte Carlo simulation:*
   - *Random sampling:* Generates numerous random scenarios for input parameters like renewable energy generation, load demand, and component failure rates [93].
   - *System simulation:* Simulates the HRES performance for each scenario, considering the energy balance, storage utilization, and system reliability [95].
   - *Statistical analysis:* Analyzes the simulation results to determine the probability of system failure, energy deficit, and other performance metrics [113].

2. *Stochastic optimization:*
   - *Incorporates uncertainty:* Formulates the optimization problem with uncertain parameters, such as renewable energy generation and load demand [114].
   - *Solves for robust solutions:* Determines optimal component sizes that minimize costs while ensuring system reliability under various uncertain conditions [115].
   - *Other techniques:* Stochastic dynamic programming, stochastic linear programming, and robust optimization [116].

3. *Markov chain models (MCMs):*
   These models are a powerful tool for sizing smart grid systems. They are particularly useful for modeling the stochastic nature of RESs, such as solar and wind power. They are used as discrete-time Markov chains or continuous-time Markov chains [117,118].

1.5.9.4   Metaheuristic optimization algorithms

This technique can be used to determine the optimal tariff in optimal operation or in optimal sizing of smart grid components [25,31]. A detailed description of these optimization algorithms is shown in section 1.9.

1.6   WLF in smart grids

WLF is a critical tool for accurate smart grid operations. It involves the accurate prediction of both weather conditions and electricity demand, which are essential for ensuring the reliable, efficient, and sustainable operation of power systems [25,31].

1.6.1   Benefits of WLF

The benefits of considering the WLF tool in the operation and design of the smart grid systems are listed in the following points:

- *Enhanced grid reliability:* Accurate forecasts enable grid operators to anticipate changes in energy demand and generation, allowing them to adjust operations to maintain grid stability. By forecasting abnormal operating conditions, smart grid operators can adjust the DR programs to reduce load and prevent overloading the grid. Additionally, accurate forecasts of renewable energy generation can help smart grid operators integrate these variable resources more effectively, reducing the reliance on traditional fossil fuel-based power plants and improving overall grid resilience.

- *Optimized energy dispatch:* By predicting future load and renewable energy generation, smart grid operators can optimize the dispatch of conventional power plants (if it is included) and/or ESSs, reducing operational costs. This involves scheduling the operation of power plants to meet the predicted demand while minimizing fuel consumption and emissions. Additionally, ESSs can be strategically charged and discharged to balance supply and demand, improving grid stability and reducing the need for peak power plants.
- *Improved renewable energy integration:* WLF can help integrate RESs like solar and wind power, which are highly dependent on weather conditions, into the grid more efficiently. By accurately predicting renewable energy generation, grid operators can adjust the dispatch of conventional power plants and ESSs to balance supply and demand, reducing the risk of grid instability and curtailment of renewable energy. Additionally, WLF can be used to optimize the operation of renewable energy systems, such as sizing ESSs and scheduling maintenance activities.

Despite the great benefits of the use of WLF in the operation and sizing of the smart grid systems, it has many challenges such as

- *Weather variability:* Weather conditions are inherently unpredictable, making accurate weather forecasts challenging.
- *Load uncertainty:* Human behavior, economic factors, and technological advancements can significantly impact electricity demand, making LF complex.
- *Data quality and availability:* Reliable and high-quality data is essential for accurate forecasting, but data quality and availability can vary across different regions.

1.6.2   Advanced techniques for WLF

WLF are critical components of modern energy systems. Accurate forecasts enable utilities to optimize grid operations, improve energy efficiency, and ensure reliable power delivery. Traditional forecasting methods, such as statistical models and physical models, have been widely used, but recent advancements in data science and ML have opened up new possibilities for more accurate and robust forecasting.

This delves into advanced techniques for WLF, exploring the latest methodologies and their applications. We will discuss the challenges and opportunities in the field, highlighting the importance of accurate and reliable forecasts for the energy sector.

1.6.2.1   Statistical methods

Traditional statistical methods, such as time series analysis, ARIMA, and exponential smoothing, have been foundational tools for WLF. However, recent advancements have led to the development of more sophisticated techniques that can capture the complex dynamics of energy consumption patterns such as:

1. *State space models*

    State space models offer a powerful framework for modeling the underlying dynamics of a system. These models represent the system's state using a set of latent variables that evolve over time. By incorporating a wide range of factors, including weather conditions, economic indicators, and policy changes, state space models can provide more accurate forecasts. State space model equations are shown in the following:

    - State equation as shown in (5): This equation essentially states that the next state of the system, $x(k+1)$, is a function of its current state, $x(k)$, the input applied to the system, $u(k)$, and a stochastic process, $w(k)$, which accounts for uncertainties and disturbances.
    - The state vector, $x(k)$, encapsulates the system's internal state at time step k. It can represent various physical quantities, such as the position and velocity of a mechanical system, the voltage and current in an electrical circuit, or the temperature and pressure in a chemical process.
    - The state transition matrix, A, determines how the system's state evolves over time in the absence of any external input. The input matrix, B, maps the input vector, $u(k)$, to the state vector, determining how the input affects the system's state.
    - The process noise, $w(k)$, accounts for uncertainties and disturbances that may affect the system's behavior. It is often modeled as a random variable with a known probability distribution, such as Gaussian noise

    $$x(k+1) = Ax(k) + Bu(k) + w(k) \qquad (1.5)$$

    where
    - ○ $x(k)$: State vector at time t
    - ○ A: State transition matrix
    - ○ B: Input matrix
    - ○ $u(k)$: Input vector at time step k.
    - ○ $w(k)$: Process noise, often assumed to be Gaussian white noise.

    - **Observation equation:** The observation equation, also known as the measurement equation, relates the system's internal state to the external measurements or observations. It describes how the system's state is observed through noisy measurements. Mathematically, it can be represented as

    $$y(k) = C\,x(k) + v(k) \qquad (1.6)$$

    where
    - ○ $y(k)$: Observation vector at time t
    - ○ C: Observation matrix
    - ○ $V(k)$: Observation noise, often assumed to be Gaussian white noise

The observation matrix, C, maps the state vector, $x(k)$, to the observation vector, $y(k)$. It determines which components of the state vector are directly observable and how they are combined to form the measurement. The measurement noise, $v(k)$, accounts for the errors and uncertainties associated with the measurement process. It is often modeled as a random variable with a known probability distribution, such as Gaussian noise. By combining the state equation and the observation equation, we can form a complete state space model. This model can be used to estimate the system's state, predict future behavior, and design control strategies.

2. *Kalman filtering:*

A key technique within state space models, Kalman filtering, enables real-time estimation of the system's state, making it suitable for online forecasting and adaptive control. It involves two main steps: prediction and update. In the prediction step, the Kalman filter predicts the system's state at the current time step based on the previous state estimate and the system dynamics. Meanwhile, the update step in Kalman filtering incorporates new measurement data to refine the predicted state estimate. This step involves calculating the Kalman gain, updating the state estimate, and updating the covariance matrix.

3. *Bayesian methods*

Bayesian methods provide a probabilistic approach to inference, allowing for the incorporation of prior knowledge and uncertainty into the forecasting process. By leveraging Bayesian techniques, it is possible to obtain more robust and reliable forecasts, especially when dealing with limited data.

- *Hierarchical Bayesian models:* These models can account for hierarchical structures in data, such as regional and seasonal variations, to improve forecasting accuracy.
- *Bayesian structural time series models:* These models combine time series components (trend, seasonal, and cyclical) with structural components (regressors, interventions) to provide flexible and interpretable forecasts.

4. *Transfer learning*

Transfer learning is a technique that leverages knowledge gained from one domain (e.g., a region with abundant data) to improve the performance of models in another domain with limited data. By adapting models to new domains, it is possible to improve the accuracy of WLF forecasts, particularly in regions with limited historical data.

- *Domain adaptation:* Techniques like feature distribution alignment can be used to bridge the gap between source and target domains.
- *Multi-task learning:* A single model can be trained to perform multiple related tasks simultaneously, such as forecasting load in different regions or predicting multiple load components.

1.6.2.2   Machine learning

ML techniques have revolutionized WLF by enabling the discovery of complex patterns and non-linear relationships in data. Several techniques used ML have been used in WLF of smart grid systems such as:

- *Neural networks:* Neural networks, particularly deep learning architectures, have emerged as powerful tools for WLF in smart grid applications. These models excel at capturing complex patterns and dependencies within large datasets, making them well-suited for predicting intricate phenomena like weather patterns and energy consumption.
- *Recurrent neural networks (RNNs):* RNNs are well-suited for sequential data, such as time series data. They can capture long-term dependencies and temporal patterns in load data. However, they can suffer from the vanishing gradient problem, limiting their ability to learn long-term dependencies.
- *Long short-term memory (LSTM) networks:* LSTMs are a type of RNN that addresses the vanishing gradient problem by using a memory cell to store information over long time intervals. They are particularly effective for complex time series forecasting. LSTMs have a more complex structure involving input, output, and forget gates that control the flow of information through the network.
- *Support vector regression (SVR):* SVR is a powerful technique for regression problems, capable of handling high-dimensional data and non-linear relationships. It aims to find an optimal hyperplane that maximizes the margin between the hyperplane and the data points.
- *Random forests:* Random forests are ensemble learning methods that combine multiple decision trees to improve prediction accuracy and reduce overfitting. Each tree in the forest is trained on a random subset of the data and features, leading to diverse and robust models.
- *Gradient boosting machines (GBM):* GBM is an ensemble learning technique that iteratively builds models to minimize prediction errors, leading to highly accurate forecasts. It works by sequentially adding trees to the model, with each new tree focusing on correcting the errors of the previous trees.

1.6.2.3   Hybrid models

Combining the strengths of statistical and ML techniques can further enhance WLF accuracy.

- *Statistical-ML hybrids:* Integrating statistical models with ML algorithms can improve the interpretability and robustness of forecasts.
- *Hybrid ARIMA-neural network model:* One popular hybrid approach combines ARIMA models with neural networks. The ARIMA model can capture the linear trends and seasonal patterns in the data, while the neural network can learn complex non-linear relationships.
- *Hybrid Bayesian and ML model:* Bayesian methods can be combined with ML techniques to incorporate prior knowledge and uncertainty into the model. For

example, a Bayesian neural network can be used to learn the parameters of a neural network in a probabilistic framework.

- *Ensemble methods:* Ensemble methods combine multiple models to improve overall performance and reduce variance.

1.6.2.4 Data-driven approaches

Data-driven approaches have revolutionized WLF in smart grid applications. By leveraging large amounts of historical data and advanced machine-learning techniques, these methods can provide highly accurate and reliable forecasts.

1.7 EVI in smart grids

The rapid electrification of transportation, driven by the increasing adoption of EVs (EVs), is reshaping the energy landscape. As EVs become more prevalent, their integration with the electric grid presents both opportunities and challenges. EVI into smart grids offers a promising avenue to enhance grid resilience, optimize energy utilization, and accelerate the transition to a sustainable energy future.

The rapid increase in EV adoption has introduced new challenges to the electric grid. Uncoordinated charging of EVs can strain the grid, leading to potential instability and increased peak demand. However, by coordinating EV charging with periods of high renewable energy generation, the negative impacts of EVs on the grid can be mitigated. The coordination can be implemented indirectly by a suitable DR strategy such as ToU or RTP. By increasing electricity tariffs during periods of high demand or low renewable energy generation and decreasing them during periods of low demand or high renewable energy generation, utilities can incentivize EV owners to shift their charging to off-peak hours. This can help balance the grid, reduce peak demand, and improve grid stability. This approach, known as G2V technology, can be implemented using automatic EVA that makes decisions on EVs charging based on dynamic electricity tariffs. By incentivizing EV owners to charge their vehicles during periods of low demand or high renewable energy generation, G2V can help balance the grid, reduce peak demand, and improve overall grid stability. Studies have shown that coordinated charging of the G2V can significantly reduce charging costs for EV owners and enhance the reliability of the smart grid. For instance, one study reported a 34% reduction in total charging costs when using an EV aggregator with demand response (G2V) compared to unregulated charging [76].

The increasing adoption of EVs presents a unique opportunity to enhance grid resilience and integrate RESs. By leveraging V2G technology, EVs can function as both energy consumers and suppliers, offering a flexible and efficient solution for EM. During periods of high renewable energy generation, such as during sunny days or windy nights, EVs can store excess energy in their batteries. Subsequently, this stored energy can be discharged back to the grid during periods of peak demand or low renewable energy generation, helping to balance the grid and reduce the reliance on fossil fuel-based power plants. Furthermore, by strategically scheduling EV charging and discharging based on DR pricing, EV owners can reduce their electricity costs. By charging their vehicles during off-peak hours when electricity prices are lower, EV owners can significantly reduce their energy bills. Additionally, V2G technology can enable EVs to provide ancillary services to the grid, such as frequency regulation and voltage control, further enhancing grid stability and reliability. However, the widespread adoption of V2G technology faces several challenges, including the limited battery capacity of EVs, the fast degradation of EV batteries, the need for advanced charging infrastructure, and the development of robust communication networks. Overcoming these challenges requires a coordinated effort from policymakers, utilities, and EV manufacturers to create a supportive regulatory and technological environment. Despite these challenges, the potential benefits of V2G technology are significant. By integrating EVs into the grid as flexible energy resources, we can move towards a more sustainable and resilient energy future. One of the primary challenges hindering the widespread adoption of V2G technology is public acceptance. Many EV owners are concerned about the potential negative impact of V2G on battery life, fearing that frequent charging and discharging cycles could accelerate battery degradation. To address these concerns, it is crucial to educate the public about the benefits of V2G, such as reduced electricity costs and improved grid reliability. Additionally, advancements in battery technology and intelligent charging algorithms can help mitigate battery degradation and ensure the long-term viability of V2G. This important issue is discussed deeply in [77] where it shows higher degradation of the batteries associated with the V2G than the G2V technology, but this increase in degradation is much lower than the benefits gained from the participation in the V2G technology. Furthermore, to address these challenges and fully realize the potential of EVI, a comprehensive approach is required, involving collaboration between utilities, policymakers, EV manufacturers, and consumers. By developing innovative technologies, implementing supportive policies, and educating the public, we can unlock the full potential of EVI and create a more sustainable and resilient energy future. Moreover, to encourage EV owners to participate in V2G programs, it is crucial to address concerns about battery degradation and economic viability. By developing advanced battery management systems and offering attractive financial incentives, utilities, and policymakers can promote the adoption of V2G technology and contribute to a more sustainable and resilient energy future.

Several studies have explored the benefits of V2G technology and how it can be leveraged to coordinate EV charging and discharging based on DR strategies [77,102]. These studies have demonstrated that V2G can significantly reduce electricity costs for EV owners and enhance grid stability. By strategically scheduling charging and discharging periods, EV owners can earn additional revenue through grid services. For instance, one study showed that EV owners could potentially earn a net profit of $3245 per year by participating in V2G programs, even after accounting for operating costs. These findings highlight the economic and environmental benefits of V2G technology and its potential to revolutionize the way we interact with the electric grid.

By integrating EVs into smart grids through V2G technology, several benefits can be realized as shown in the following points:

- *Peak demand reduction:* By strategically scheduling EV charging during off-peak hours, utilities can reduce peak demand, easing the strain on the grid infrastructure.
- *Enhanced grid reliability:* EVs can serve as distributed energy storage resources, providing ancillary services like frequency regulation and voltage control.
- *Increased renewable energy integration:* EVs can absorb excess renewable energy, especially during periods of low demand, promoting the integration of RESs.
- *Reduced emissions:* The widespread adoption of EVs can significantly reduce GHG emissions and improve air quality.

A significant portion of EV usage, estimated to be between 5% and 95% of the time, involves the vehicle being parked [119]. During these periods, an EVA can optimize charging and discharging schedules to maximize benefits for EV owners, such as reduced electricity costs and potential revenue from V2G or G2V participation. However, a major barrier to widespread V2G adoption is EV owners' skepticism about the potential impact on battery life and the economic viability of such programs. To address these concerns, accurate battery wear and economic models are crucial to assure EV owners of the potential benefits and minimize their worries.

While unidirectional G2V technology offers clear benefits to EV owners, such as reduced charging costs, it is important to consider the potential impact on battery life and the overall economic viability of V2G. While some studies have focused on the immediate benefits of G2V, a more comprehensive analysis should account for the long-term effects of battery degradation and the potential for additional revenue through V2G services [120–122]. By developing accurate battery wear models and conducting detailed economic assessments, it is possible to optimize the utilization of V2G technology and maximize its benefits for both EV owners and grid operators.

The price differential between peak and off-peak electricity rates is a critical factor in the viability of V2G applications. This price spread must be sufficient to compensate for energy losses during charging and discharging cycles, as well as the accelerated battery degradation associated with frequent cycling. Several studies have investigated the minimum price differential required to incentivize EV owners to participate in V2G programs in specific markets, such as the UK, Germany, and Sweden [123].

While many studies advocate for the benefits of V2G technology, some research suggests that its economic viability and technical feasibility may be limited [124]. Concerns have been raised about the potential for accelerated battery degradation and the need for significant investments in charging infrastructure. To address these concerns, it is essential to develop accurate battery wear models and conduct comprehensive economic analyses to evaluate the long-term costs and benefits of V2G [125]. A new study aims to contribute to this ongoing debate by providing a detailed analysis of EV battery wear under various operating conditions is introduced in [76].

Several studies have investigated optimal charging strategies for EVs to minimize battery degradation and maximize economic benefits. These studies have explored various techniques, including:

- *Battery wear modeling:* Researchers have developed sophisticated battery wear models that consider factors such as temperature, state-of-charge (SoC), depth of discharge (DoD), and charging/discharging rates. These models help assess the long-term impact of V2G operations on battery life.
- *Optimal charging scheduling:* Various optimization techniques, such as dynamic programming and ML, have been employed to develop intelligent charging strategies that minimize battery degradation while maximizing energy efficiency and economic benefits.
- *Real-time pricing and demand response:* By utilizing RTP signals and demand response programs, EV owners can be incentivized to charge their vehicles during off-peak hours or periods of high renewable energy generation, further reducing wear and tear on batteries.

These studies have provided valuable insights into the potential of V2G technology and the importance of careful battery management. By addressing concerns about battery degradation and economic viability, these research efforts can help accelerate the adoption of V2G and contribute to a more sustainable and resilient energy future.

The Society of Automotive Engineers (SAE) has defined three primary levels of EV charging based on the J177 standard [126]:

- Level 1 charging: This is the slowest charging method, typically using a standard 120-V AC outlet. It can take 12-24 h to fully charge an EV battery.
- Level 2 charging: This is a faster charging method, using a 240-V AC outlet. It typically takes 4-8 h to fully charge an EV battery.
- Level 3 charging (DC fast charging): This is the fastest charging method, using direct current (DC) power to rapidly charge an EV battery. It can add significant range in as little as 20-30 min.

It is important to note that faster charging methods, such as Level 3, can potentially accelerate battery degradation due to the higher charging rates and increased thermal stress. Therefore, while faster charging is convenient, it may have a slight negative impact on battery longevity [127].

1.7.1 Novel EV battery's wear model

This study introduces a novel incremental aging model for EV batteries, addressing a key challenge in accurately predicting battery degradation. Previous research often relied on simplified models with unknown parameters, requiring extensive testing and calibration. In contrast, this approach leverages publicly available data on the relationship between achievable cycle count (ACC) and DoD to estimate battery wear. By considering the instantaneous changes in SoC, DoD, and temperature, the proposed model provides a more accurate representation of real-world battery aging. This improved understanding of battery degradation can help optimize V2G strategies and ensure the long-term viability of EV batteries.

The relation between the *ACC-DoD* characteristics can be modeled from the characteristic curve shown in Figure 1.11 as shown in (1.7) [76]

$$ACC = aD^{-b} \tag{1.7}$$

where D is the *DoD* which can be obtained from (1.8), and a and b are the parameters that can be determined from the *ACC-DoD* characteristics.

The wear of the EV battery with constant power from t_1 to $t_1+\Delta t$ can be determined from the following equation [76]:

$$W(P,S) = \frac{K_W \cdot P_B^2}{b \cdot E_B^R}\left[\left(1 - S_{\text{int}} + \frac{P_B \cdot \Delta t}{E_B^R}\right)^b - (1 - S_{\text{int}})^b\right] \tag{1.8}$$

where $K_W = \frac{1}{\eta_B^2 \cdot a \cdot E_B^R \cdot (1+\theta)}$ in which θ is the ratio of the capacity that should replace the EV battery when it reaches it (80% in most studies [76]), E_B and E_B^R are the current and rated energy of the EV battery, respectively, P_B is the charging/discharging power from the battery, and S_{int} is the initial SoC of each time window (h).

For this study, a conversion of 0.18 kWh/mile was chosen, which is similar to the measured fuel economy of a Nissan leaf, to translate distance traveled into an energy expenditure to anticipate the energy consumption of the EV [76].

A significant portion of daily commutes, estimated to be around 80%, typically cover distances between 60 and 70 km [76]. Given that EVs are plugged in for approximately 90% of the day [128], there is a significant opportunity to optimize their charging and discharging schedules. Uncoordinated charging, particularly high-power charging, can lead to a 60% increase in peak power demand [129].

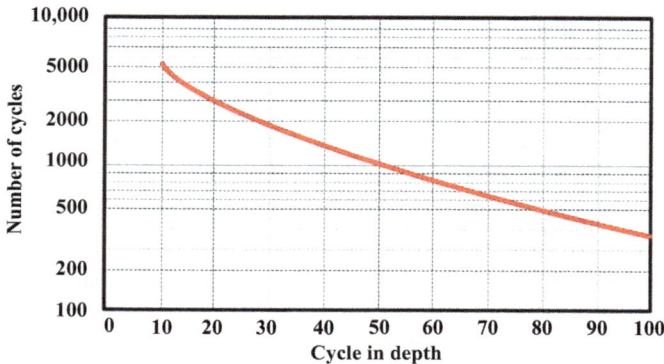

Figure 1.11   The ACC along with DoD for LIB used in this study

By implementing dynamic pricing strategies, such as TOU or RTP, grid operators can encourage EV owners to charge their vehicles during off-peak hours, reducing peak demand and improving grid stability. Furthermore, V2G technologies can enable EVs to provide ancillary services to the grid, further enhancing grid resilience and reducing emissions.

A log-normal distribution function, as shown in (1.9) [130], can be employed to model the daily driving distance of EVs. Figure 1.12 visually represents the probability density function of this distribution.

$$f_{des}(L_{EV}) = \frac{1}{L_{EV}\sigma_{EV}\sqrt{2\pi}} e^{\left[-\frac{(\ln -\mu_{EV})^2}{2\sigma_{EV}^2} \right]} \tag{1.9}$$

where L_{EV} is the daily trip distance, σ_{EV} is the average daily distance of EV, and μ_{EV} is the variance of the daily distance of EV.

1.7.2 Simulation program

A MATLAB-based simulation model was developed to implement the proposed charging and discharging strategies Figure 1.13. As shown in Figure 1.14, the simulation process begins by generating random driving distances and departure times based on stochastic models. The EV model then simulates the vehicle's energy consumption during the trip, considering factors such as driving conditions and vehicle efficiency.

The optimized charging/discharging schedule, generated by the optimization algorithm, is used to calculate the hourly energy flow into and out of the EV battery. This information, along with the battery's ACC-DoD characteristics, is fed into the battery wear model to estimate the hourly and daily wear. The economic model then assesses the charging costs, battery degradation costs, and potential revenue from V2G services.

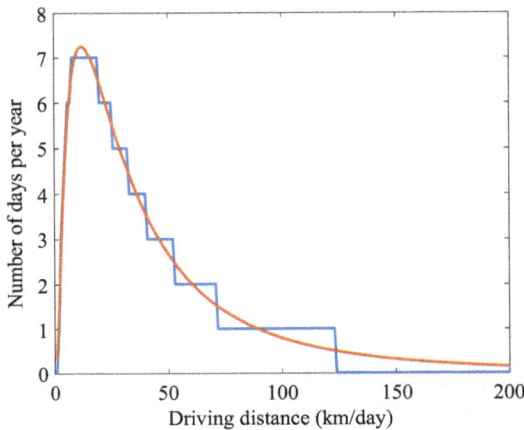

Figure 1.12   The daily driving distance distribution of the EV for a full year of operation

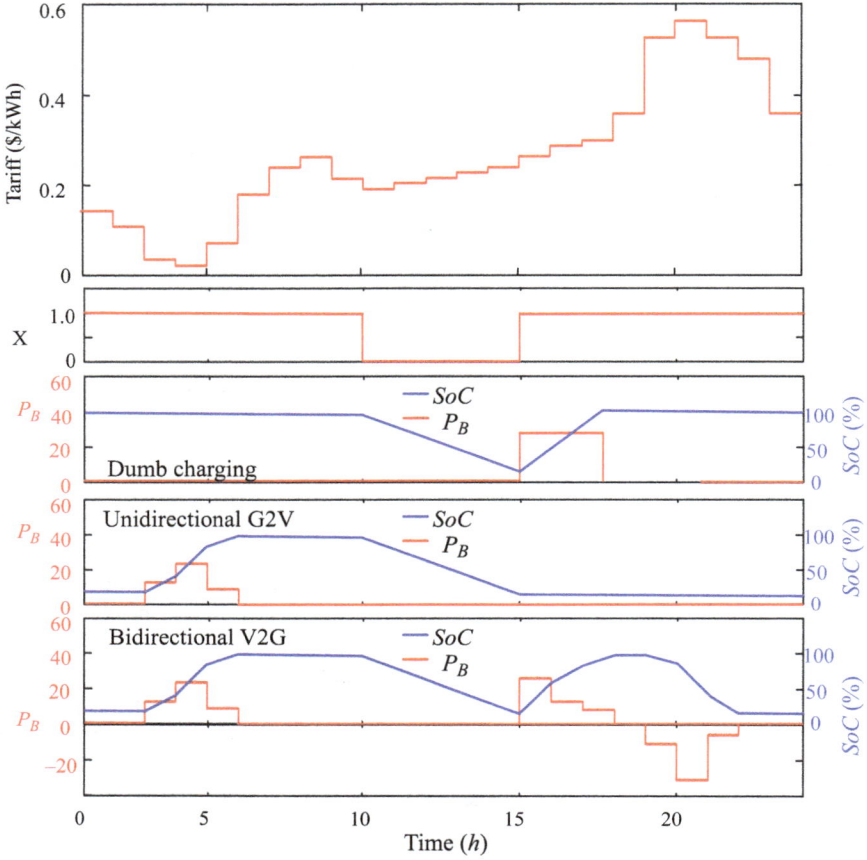

Figure 1.13   The charging/discharging schedules of different technologies under study

Based on the economic analysis, the optimization algorithm iteratively adjusts the charging/discharging schedule to minimize costs and maximize profits. This iterative process continues until a satisfactory solution is achieved. A detailed explanation of each component of the simulation model is provided in the following sections.

1.7.3   Simulation results

Unregulated charging is charging the battery with its maximal charging power once it is attached to the supply means. This technology does not schedule the charging as shown in Figure 1.15. The simulation of this technology is carried out here for comparison. The charging cost of the EV under study with this technology is $917.8 [76].

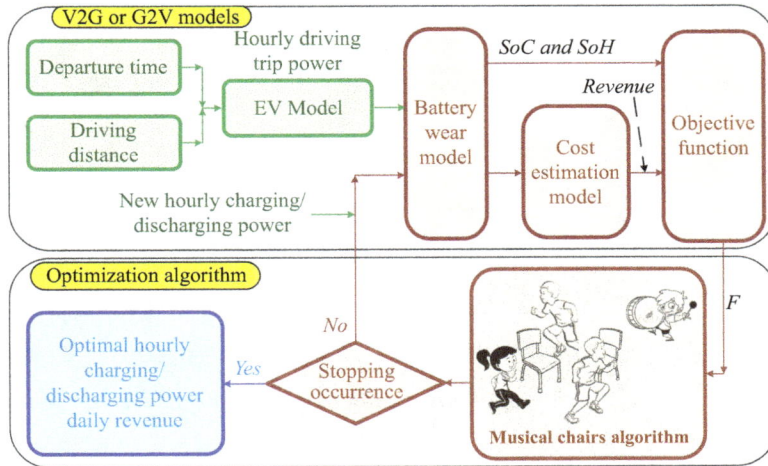

Figure 1.14   The block diagram of the simulation program used to implement the proposed strategy

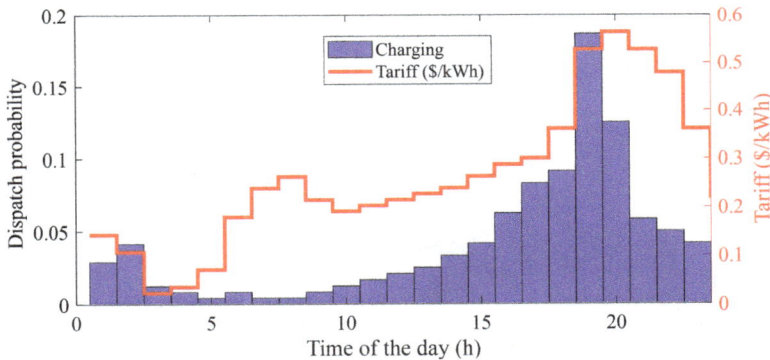

Figure 1.15   The dispatch probability during day hours for a complete year along with the tariff ($/kWh) for unregulated charge technology

The optimization algorithm suggests the charging power and determines the corresponding cost for these suggestions. The optimization algorithms vary the scheduling of the charging/discharging power again searching for the lowest daily cost. The lowest daily cost for sure occurs in the lowest tariff as shown in Figure 1.16. The analysis reveals that the optimal charging strategy, enabled by the G2V technology, concentrates charging activity during the lowest tariff period, specifically around 3:00 AM. While there are minor deviations due to real-world driving patterns and optimization algorithm adjustments, the majority of charging occurs during these off-peak hours. This strategic approach significantly reduces the annual battery degradation rate to 2.43%, which is 95.3% of the wear experienced with

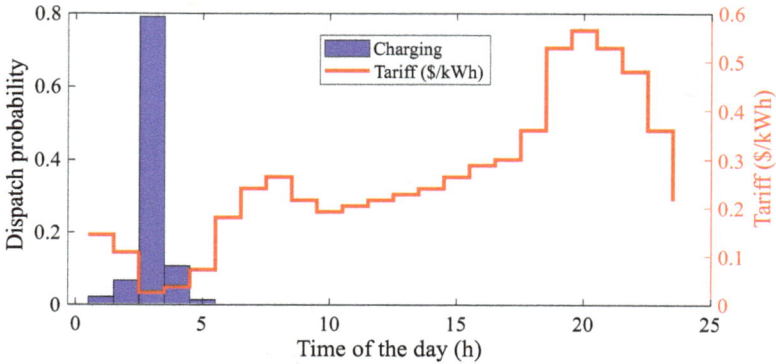

Figure 1.16   The dispatch probability during day hours for a complete year along with the tariff ($/kWh) for G2V technology

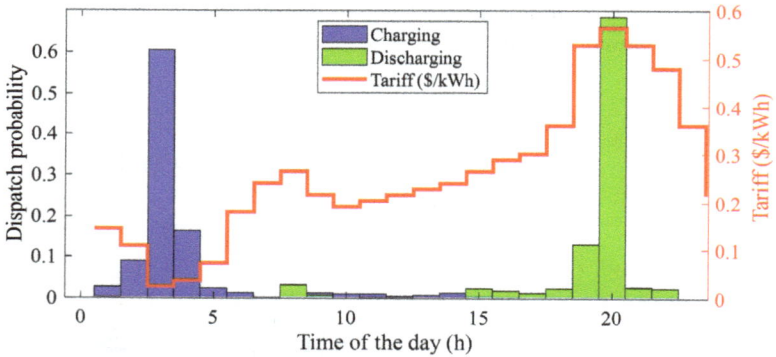

Figure 1.17   The dispatch probability during day hours for a complete year along with the tariff ($/kWh) for V2G technology

unregulated charging (UC). Consequently, the battery lifespan is extended to 8.23 years, a 4.7% improvement over UC. The most substantial benefit of G2V lies in reduced charging costs. The annual charging cost with G2V is only $436.6, compared to $917.8 for UC, representing a 47.5% reduction. This reduction, coupled with a 5% decrease in annual battery wear costs, results in a total yearly cost of $996.2 for G2V, compared to $1505.3 for UC. This translates to a 34% reduction in overall costs, demonstrating the significant economic advantages of V2G technology.

The V2G bidirectional charging/discharging technology can substantially reduce the charging cost by charging the EV at periods with low tariffs and discharging the saved energy at times of high tariffs and getting benefits from the tariff difference. The charging/discharging schedule for V2G technology for a certain day as an example is illustrated in Figure 1.17. The simulation results demonstrate that the V2G strategy effectively leverages ToU pricing to optimize charging and discharging schedules. The EV is primarily charged during off-peak hours (4:00 AM to 6:00 AM) to minimize costs. When the EV is required for a trip (10:00 AM to 15:00 PM), it discharges its stored energy to meet the vehicle's needs. Subsequently, during peak hours (19:00 PM to 21:00 PM), the EV discharges excess energy back to the grid, generating additional revenue. However, the V2G strategy also results in increased battery wear due to frequent charging and discharging cycles. As shown in Table 1.8, the annual battery wear for V2G is significantly higher than that of UC and G2V. This increased wear reduces the battery's lifespan to 3.7453 years, compared to 7.8431 and 8.23 years for UC and G2V, respectively. While the V2G strategy incurs higher charging and battery wear costs, it also generates substantial revenue from selling energy back to the grid. The total annual cost for V2G is $1996.3, which is higher than the costs of UC and G2V. However, the additional revenue of $5240.7 from V2G services results in a net profit of $4749.7, surpassing the total costs of UC and G2V.

The simulation results demonstrate the significant benefits of V2G technology compared to UC and G2V strategies. While V2G leads to slightly higher battery degradation due to increased charging and discharging cycles, it offers substantial economic advantages listed in the following points:

• Reduced charging costs: V2G can significantly reduce annual charging costs, as seen in the 33% reduction compared to UC.
• Increased revenue: By selling energy back to the grid during peak hours, V2G can generate additional revenue for EV owners, leading to a net profit.
• Improved grid stability: V2G can contribute to grid stability by providing flexibility and supporting peak shaving and valley filling.

While V2G offers numerous advantages, it is crucial to address concerns about battery degradation. By optimizing charging and discharging schedules and employing advanced battery management techniques, it is possible to mitigate the impact of V2G on battery life.

1.8 IP_RESs in smart grids

The global shift toward sustainable and clean energy sources has led to a significant surge in the adoption of RESs like solar, wind, and hydro-power. These renewable

Table 1.8   The comparison between three different technologies under study

Items	Dumb charge	G2V	V2G
Yearly battery Degradation (%)	2.55	2.43	5.34
Battery life span (Years)	7.8431	8.23	3.7453
Yearly charging cost ($)	917.8	436.6	765.4
Yearly wear cost ($)	587.5	559.6	1230.9
Total yearly cost ($)	1505.3	996.2	1996.3
Income due to V2G ($)	–	–	5240.7
Yearly revenue compared to UC ($)	–	508.7	4749.7

sources offer numerous benefits, including reduced GHG emissions, reduced dependence on fossil fuels, and increased energy security. However, the integration of RESs into the traditional power grid presents a myriad of challenges, necessitating the evolution of the grid into a smarter, more flexible, and resilient system. Smart grids, equipped with advanced technologies such as intelligent sensors, communication networks, and automation systems, provide the platform for seamless integration of RESs. By enabling real-time monitoring, control, and optimization of energy flows, smart grids can accommodate the intermittent and fluctuating nature of renewable energy generation. Moreover, they can enhance grid reliability, improve energy efficiency, and facilitate the emergence of new energy services, such as demand response and EV charging. The IP_RESs in smart grids offers several advantages discussed above, and one of these advantages is the increase in the penetration of RESs to power systems. Moreover, smart grid technology can upgrade existing grid infrastructure to accommodate the IP_RESs may require significant investment [102].

To increase the penetration of RESs in smart grids, a multi-faceted approach is necessary. This involves technological advancements, policy support, and consumer engagement. Here are some key strategies.

1.8.1   Technological advancements

- *Advanced ESSs:* Developing efficient and cost-effective energy storage solutions, such as batteries, pumped hydro storage, and compressed air energy storage, can help balance the intermittent nature of renewable energy generation.
- *Smart grid technologies:* Implementing AMI and intelligent grid control systems can optimize energy distribution and integrate RESs more seamlessly.
- *Power electronics:* Developing high-efficiency power electronic converters can improve the performance and reliability of renewable energy systems.

1.8.2   Policy and regulatory support

- *Renewable energy targets:* Setting ambitious renewable energy targets and providing clear policy frameworks can incentivize investment in RES projects.
- *Feed-in tariffs and subsidies:* Offering financial incentives can make renewable energy projects more economically viable.
- *Net metering:* Enabling consumers to generate their own electricity and sell excess power back to the grid can encourage rooftop solar installations.
- *Grid modernization:* Investing in grid infrastructure upgrades to accommodate the increasing penetration of RESs.

1.8.3   Consumer engagement

- *Education and awareness:* Raising public awareness about the benefits of renewable energy and energy efficiency can encourage consumer adoption.

- *Community solar programs:* Allowing individuals to invest in local solar projects can make renewable energy more accessible.
- *Time-of-use pricing:* Implementing ToU pricing can incentivize consumers to shift their energy consumption to off-peak hours when renewable energy generation is abundant.

1.8.4 Grid integration challenges and solutions

- *Intermittency and variability:* Advanced forecasting techniques and ESSs can help mitigate the impact of intermittent renewable energy generation.
- *Voltage fluctuations:* Smart inverters with voltage control capabilities can help regulate voltage levels.
- *Grid stability:* Grid operators can implement measures like reactive power control and frequency regulation to maintain grid stability.

By addressing these challenges and implementing appropriate strategies, we can accelerate the transition to a sustainable and low-carbon energy future. Several studies have been introduced in the literature to evaluate the IP_RESs due to the use of smart grid concepts such as [19,131,132].

1.9 Optimization algorithms

AI techniques have emerged as powerful tools for optimizing the design and sizing of HRES [9,13,18,22,24,25,30,31,41,42,76–78]. These techniques can handle the complexities of HRES, including the variability of RESs, uncertain load demands, and the need to balance economic and environmental factors. The AI optimization algorithms used for optimal operation and sizing in smart grid applications are shown in Figure 1.18.

1.9.1 Machine learning

- *Predictive modeling:* ML models can accurately predict future energy demand and renewable energy generation, enabling better sizing decisions.
- *Feature engineering:* By carefully selecting and engineering relevant features, such as weather data, historical load profiles, and economic indicators, ML models can improve their predictive accuracy.
- *Time series analysis:* Techniques like ARIMA, LSTM, and GRU can effectively capture the temporal dependencies in energy data, leading to more accurate forecasts.

1.9.2 Metaheuristic optimization algorithms

- *Genetic algorithms:* Inspired by natural selection, these algorithms can explore a vast solution space to find optimal solutions.
- *Particle swarm optimization:* Simulates the social behavior of bird flocks to find optimal solutions.

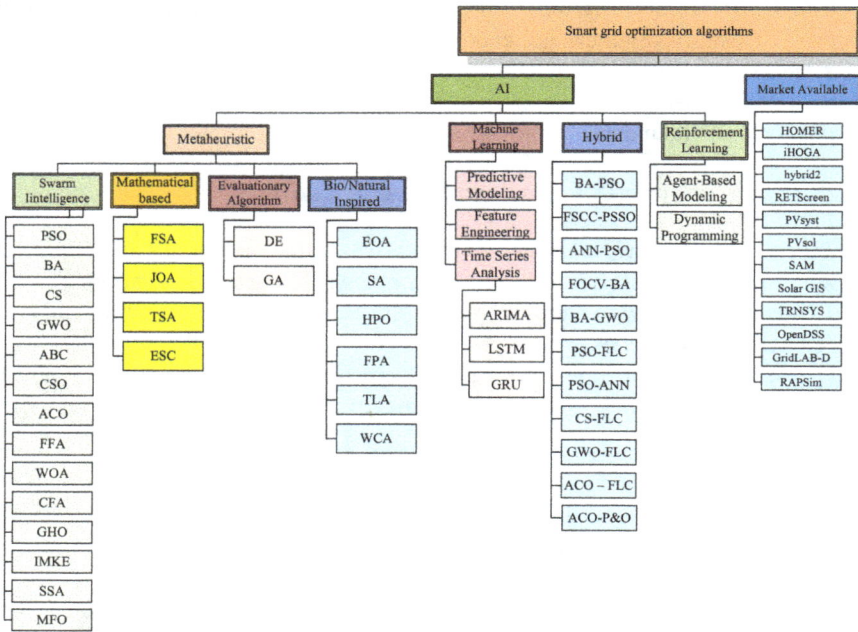

Figure 1.18   Classifications of PV MPPT techniques

- *Differential evolution:* A population-based optimization algorithm that uses mutation and crossover operators to evolve solutions.

1.9.3   Reinforcement learning

- *Agent-based modeling:* Agents can learn to make optimal decisions about energy generation, storage, and distribution through trial and error.
- *Dynamic programming:* Can be used to find optimal policies for EM and control.

1.9.4   Hybrid approaches

- *Combining AI and analytical techniques:* Hybrid approaches can leverage the strengths of both worlds, combining the flexibility and adaptability of AI with the precision of analytical methods. For example, ML can be used to predict future energy demand, and then optimization techniques can be used to determine the optimal sizing of HRES components.

1.9.5   Market available software

Several software tools are available to assist in the design, optimization, and operation of HRES. A detailed comparison between these software packages is shown in Refs. [133–137] and summarized in Table 1.9.

Table 1.9 Summary of few software tools for HRES

Software	Developed by	Year	Key features	Limitations	Suitable applications	Studies
HOMER	National Renewable Energy Laboratory (NREL)	1990s	Comprehensive analysis of hybrid power systems, including economic and environmental factors	Complex interface requires significant computational resources	Off-grid and grid-connected systems, microgrids, remote area electrification	[138–141]
iHOGA	University of Stuttgart	2000s	Specialized in optimizing the design and operation of hybrid power systems	Limited to specific system configurations and optimization algorithms	Academic research and specialized engineering applications	[141–143]
hybrid2	National Renewable Energy Laboratory (NREL)	2000s	Open-source tool for modeling and simulating hybrid power systems	Requires programming skills and customization for specific applications	Research and development, custom system design	[144,145]
RETScreen	Natural Resources Canada	1990s	Comprehensive tool for evaluating the technical and financial feasibility of renewable energy projects	Can be less flexible for complex system configurations	Feasibility studies, financial analysis, and environmental impact assessments	[146,147]
PVsyst	University of Geneva	1990s	Specialized in simulating PV systems	Limited to PV systems and their integration with other RESs	Detailed PV system design and performance analysis	[148,149]
PVsol	Valentin Software	1990s	Similar to PVsyst, focused on PV system design and simulation	Can be less flexible for complex system configurations	Detailed PV system design and performance analysis	[150,151]
SAM (System Advisor Model)	National Renewable Energy Laboratory (NREL)	2000s	Comprehensive tool for analyzing renewable energy systems, including solar, wind, and biomass	Complex interface and requires significant computational resources	Research and development, system design, and policy analysis	[147,152]
Solar GIS	Solargis	2000s	Provides solar irradiation data and other solar resource	Limited to solar resource assessment	Solar resource assessment and site selection	[153,154]

(Continues)

Table 1.9 (*Continued*)

Software	Developed by	Year	Key features	Limitations	Suitable applications	Studies
			information			
TRNSYS	University of Wisconsin-Madison	1970s	A powerful simulation tool for building energy systems, including renewable energy systems	Requires advanced modeling skills and computational resources	Detailed system simulations and energy performance analysis	[155,156]
OpenDSS	Electric Power Research Institute (EPRI)	1990s	Open-source tool for simulating power distribution systems, including those with DERs	Requires programming skills and customization for specific applications	Power system analysis, distribution system planning, and microgrid studies	[157,158]
GridLAB-D	Pacific Northwest National Laboratory (PNNL)	2000s	Open-source tool for simulating power distribution systems and microgrids	Requires programming skills and customization for specific applications	Power system analysis, distribution system planning, and microgrid studies	[159,160]
RAPSim	National Renewable Energy Laboratory (NREL)	2000s	Renewable energy resource modeling, battery storage modeling, power flow analysis, economic analysis, sensitivity analysis	Limited flexibility for custom modeling, less mature than GridLAB-D	Renewable energy system design and optimization, microgrid analysis and control	[160]
PowerFactory	DIgSILENT	1990s	A comprehensive power system analysis software that can be used for modeling and simulating smart grid components	Complex software with a steep learning curve	Power system analysis, grid planning, and operational studies	[161,162]

1.10   Advanced communication networks

The smart grid is the next-generation electric power system that is characterized by a two-way flow of electricity and information. Compared with conventional power grids that rely on one-way communication, the smart grid provides two-way communication between customers and utilities, which plays an important role in providing reliable, efficient, and secure power generation, transmission, and distribution. The smart grid will enable utilities to monitor, protect, automate, and optimize the operation of the whole electric power system from consumption to power generation. With new technologies and applications such as DERs, EVs, AMI, and home energy management systems, communication network design become very challenging as communication infra-structures cover different network segments with diverse requirements for the quality of service. In order to achieve the potential advantages of the smart grid, reliable design and implementation of secure communication infrastructures are needed to support different requirements such as reliability, bandwidth, latency, and security [163–175].

There are different architectures and models available for the smart grid. According to the National Institute of Standards and Technology (NIST), the smart grid consists of seven logical domains: bulk generation, transmission, distribution, customer, markets, operations, and service providers. The smart grid communication network represents a hierarchical architecture that interconnects among all different domains. Below, we describe a general framework for a smart grid with four main layers: physical power system layer, sensor and actuator layer, communication network layer, and management and control layer [1,173], as shown in Figure 1.19.

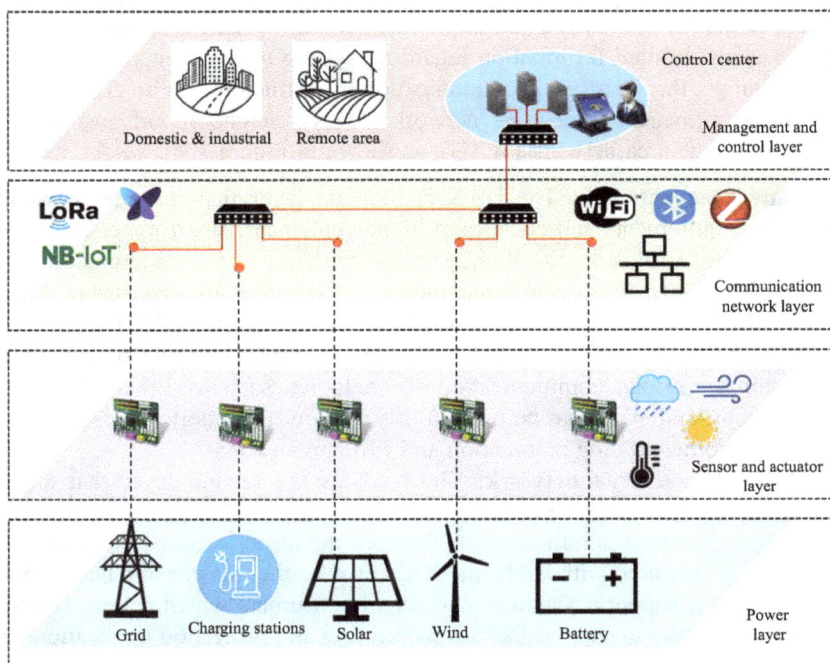

Figure 1.19   Smart grid cyber-physical system

- **Physical power system layer:** The physical layer is the main layer of the smart grid covering generation, transmission, distribution, and consumption. The main physical layer elements are generators, transmission lines, transformers, buses, and loads.
- **Sensor and actuator layer:** The sensor and actuator layer consists of smart sensors and measuring devices such as intelligent electronic devices (IEDs) and remote terminal units (RTUs), which are responsible for measuring information such as voltage, current, frequency, and status of circuit breakers. Other control devices and actuators include, for example, circuit breakers, relays, and distributed generation controllers.
- **Communication network layer:** The communication network layer consists of switches, routers, and communication medium, which is responsible for seamless data exchange between different smart grid components. The communication medium can be wired or wireless based on system requirements.
- **Management and control layer:** The control center is responsible for the control and management operation of the smart grid under different conditions. Based on received measurements, through the communication network layer, data is processed, which enables monitoring and decision-making for reliable operation.

In general, the smart grid communication infrastructure consists of a layer structure that supports data collection and control operations. There are different structures and scales for communication networks that adopt different smart grid applications such as home/building automation, smart meter reading, distribution automation, and EV integration. Both customers and power utilities may use various wired/wireless communication networks to support different applications in different domains. The design and implementation of a smart grid communication network needs detailed information regarding system requirements. Based on the coverage range, the smart grid communication infrastructure can be classified into three main segments: home area network (HAN), neighborhood area network (NAN), and wide area network (WAN), as shown in Figure 1.20.

- **Home area network:** The HAN is the first layer that manages consumer power requirements and consists of home appliances, smart meters, EVs, and RESs. The HAN is deployed in consumer premises and commercial buildings to gather information from home/building appliances and measuring devices and delivers control information for better home/building EM. Different network topologies could be configured for HAN such as start topology or mesh topology. Various communication technologies such as Ethernet, ZigBee, Bluetooth, and WiFi can be used in this domain to support different services such as home/building automation and EM applications.
- **Neighborhood area network:** The NAN is the second layer that enables information exchange between the consumer side and the power utility. Also, the NAN supports communication between the utility control center and field devices. Compared with HAN, the NAN enables the transfer of a huge amount of data as it supports a greater number of consumers which covers 100 m to 10 km. In order to support various monitoring and protection applications, the

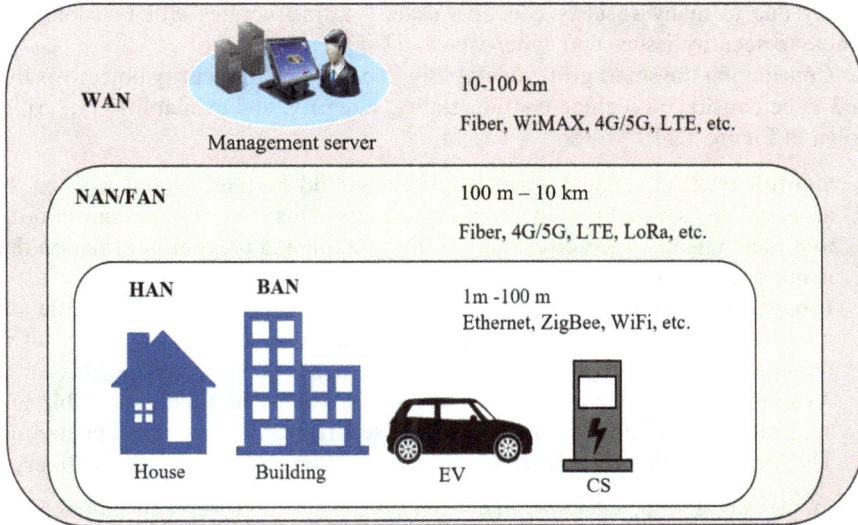

Figure 1.20   Smart grid communication network

field area network enables the information exchange among the utility control center, distribution substation, and feeder equipment using field devices such as IEDs, phasor measurement units, and RTUs. The NAN enables different smart grid applications such as smart metering, load management, demand response, and distribution automation.

- **Wide area network (WAN):** The WAN is the third layer, which serves as a backbone for the communication network between utility infrastructures and the utility control center. The WAN enables real-time monitoring and control as well as data aggregation from multiple NANs, which cover a very large area and provide real-time information about the status of the power grid. Different communication technologies such as optical fiber, LTE, 4G/5G cellular, WiMAX, and LoRa can be used to support different applications such as wide-area monitoring, wide-area protection, and wide-area control. Compared with HAN and NAN, WAN applications require long-distance coverage of 10–100 km.

1.11   Cybersecurity (CyberSec)

The smart grid can be represented as a complex cyber-physical system due to the coupling between the power infrastructures and the cyber systems, and the huge number of entities and devices connected together via the communication network. Although the advances in information and communication technologies (ICT) will enable the grid operator to support enhanced services for monitoring, controlling, and optimizing smart grid operations, cyber security must be considered with high priority due to many security concerns as ICT infrastructures will be more vulnerable to security issues and cyber-attacks [1–5].

Considering the smart grid vulnerability, the three main security objectives that need to be considered include confidentiality, integrity, and availability [1,3,5], as shown in Figure 1.21.

- **Confidentiality:** Critical smart grid data should be confidential and can be accessed and viewed only by authorized users. This is very important in order to protect unauthorized disclosure of information and prevent information that is not open to the public.
- **Integrity:** Smart grid data must maintain accuracy and consistency while any modifications or losses must be detected. The loss of integrity can result in incorrect decisions regarding power management.
- **Availability:** The smart grid data and systems must be timely available and accessible to authorized parties without security compromise when needed. The loss of availability can result in disruption of service or power delivery.

In the smart grid, customer data and smart meter power consumption are highly sensitive and must be accessed only by the utility company and customer. Data confidentiality should be protected from being acceded by unauthorized parties as customer identity and energy usage represent confidential data that can identify end-user activities.

One of the most challenging smart grid threats is false data injection, which targets data integrity. Another type of attack is called a data availability attack, which aims to delay or block the data communication where one or more sources start flooding the network and communication layer to exhaust network bandwidth, router processing capability, and servers. The cyber-attack on smart grid communication can be classified into four different attack types: device attack, data attack, privacy attack, and network availability attack [4], as shown in Figure 1.22.

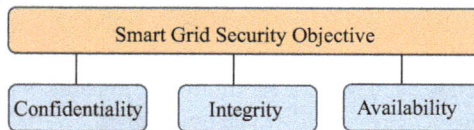

Figure 1.21   Main smart grid security objectives

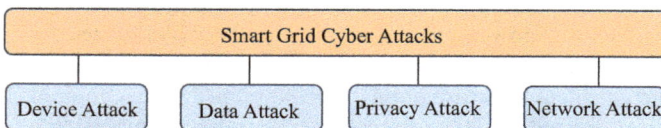

Figure 1.22   Main smart grid cyber-attacks

- *Device attack:* The device attack aims to compromise or control a grid device. In general, this is the first step toward a sophisticated attack where the compromised device can be used to perform, for example, a power outage using a compromise IED connected to a circuit breaker. Strict access control is very important to overcome device attacks.
- *Data attack:* The data attack aims to alter, insert, or delete network traffic or control information to mislead the smart grid into making wrong actions or decisions. Smart meter data, for example, can be compromised to report the wrong electricity bill. Data integrity is very important to overcome this attack.
- *Privacy attack:* The privacy attack aims to infer user privacy information such as analyzing electricity usage data. Such privacy-sensitive information must be protected from unauthorized access.
- *Network attack:* The network availability attack aims to overwhelm the communication and computational resources which results in failure or delay of the communication network. For example, man-in-the-middle-attacks compromise the integrity and availability of the power data network.

1.12 Conclusions and future work

1.12.1 Conclusions

This chapter provides a comprehensive exploration of the critical aspects of modern power systems and smart grid technologies. This in-depth analysis highlights the significant advancements and challenges associated with the transition towards a sustainable and resilient energy future. A key contribution of this chapter lies in its detailed comparison of traditional power systems and smart grids. By emphasizing the advantages and limitations of both, the chapter provides a clear understanding of the potential benefits and challenges associated with the adoption of smart grid technologies. Moreover, the chapter delves into the critical components of smart grids, such as ESSs and DSM, highlighting their crucial role in enhancing grid flexibility, reliability, and efficiency. The chapter also emphasizes the importance of accurate forecasting and optimization techniques in ensuring optimal grid operation. By analyzing various optimization techniques, including linear programming, nonlinear programming, and metaheuristic algorithms, the chapter provides valuable insights for researchers and practitioners seeking to implement effective strategies for power flow management. Furthermore, the study underscores the significance of cybersecurity in safeguarding smart grid infrastructure and sensitive data. With the increasing integration of RESs and EVs, robust cybersecurity measures are essential to mitigate potential threats and ensure the resilience of smart grids.

Finally, this chapter offers a valuable resource for researchers, policymakers, and industry professionals seeking to advance the field of smart grid technology. By providing a comprehensive overview, highlighting key challenges and opportunities, and suggesting future research directions, the chapter contributes to the ongoing efforts to build a sustainable and resilient energy future.

1.12.2 Avenues for further research

This chapter provides a solid foundation for future research in the field of smart grid technology. Several promising avenues for future investigation are identified as shown in the following points:

1. *Advanced optimization techniques and algorithms:*
   - Hybrid optimization techniques: Combining metaheuristic algorithms with traditional optimization techniques to enhance solution quality and computational efficiency.
   - Multi-objective optimization: Developing optimization frameworks that consider multiple objectives, such as minimizing costs, maximizing renewable energy integration, and improving system reliability.
   - Robust optimization: Incorporating uncertainty and variability in renewable energy generation and load demand to design robust and resilient smart grid systems.

2. *Integration of AI and ML:*
   - Advanced forecasting techniques: Developing advanced AI and ML-based forecasting models to improve the accuracy of load and renewable energy generation predictions.
   - Real-time decision making: Utilizing AI-powered decision support systems to optimize energy dispatch, demand response, and grid operations in real-time.
   - Cybersecurity applications: Employing AI and ML techniques for anomaly detection, intrusion detection, and cyberattack prevention.

3. *Innovative solutions for emerging challenges:*
   - Electric vehicle integration: Investigating strategies for integrating large-scale electric vehicle charging infrastructure into the grid, including V2G technologies.
   - Distributed energy resources: Developing advanced control and coordination mechanisms for DERs, such as distributed generation, ESSs, and demand response programs.
   - Microgrid technologies: Exploring the potential of microgrids to enhance grid resilience, reduce energy losses, and improve energy efficiency.

4. *Enhanced WLF:*
   - Incorporating advanced statistical and ML techniques to improve the accuracy and reliability of WLF.
   - Developing data-driven models that can adapt to changing environmental conditions and consumer behavior patterns.
   - Leveraging historical data, real-time sensor data, and external factors like weather forecasts to refine forecasting models.

5. *Cybersecurity and privacy:*
   - Strengthening cybersecurity measures to protect critical infrastructure from cyberattacks and data breaches.

- Developing advanced encryption techniques and intrusion detection systems to safeguard sensitive information and ensure the integrity of grid operations.
- Investigating privacy-preserving data analytics techniques to enable data-driven decision-making while protecting user privacy.

6. *Comprehensive smart grid modeling and simulation:*
   - Developing advanced simulation tools and platforms that can accurately model the behavior of complex smart grid systems.
   - Incorporating real-world data and uncertainties into simulation models to improve their predictive capabilities.
   - Utilizing simulation tools to evaluate the impact of different policies, technologies, and OS on grid performance.

By addressing these future research directions, researchers and industry practitioners can contribute to the development of more sustainable, resilient, and efficient smart grid systems, paving the way for a greener and more sustainable future.

References

[1] F. Nejabatkhah, Y. W. Li, H. Liang, and R. Reza Ahrabi, "Cyber-security of smart microgrids: A survey," *Energies*, vol. 14, no. 1, p. 27, 2020.

[2] A. Stefanov and C.-C. Liu, "ICT modeling for integrated simulation of cyber-physical power systems," in *2012 3rd IEEE PES Innovative Smart Grid Technologies Europe (ISGT Europe)*, 2012: IEEE, pp. 1–8.

[3] M. Khalaf, A. Ayad, M. H. K. Tushar, M. Kassouf, and D. Kundur, "A survey on cyber-physical security of active distribution networks in smart grids," *IEEE Access*, vol. 12, pp. 29414–29444, 2024.

[4] X. Li, X. Liang, R. Lu, X. Shen, X. Lin, and H. Zhu, "Securing smart grid: cyber attacks, countermeasures, and challenges," *IEEE Communications Magazine*, vol. 50, no. 8, pp. 38–45, 2012.

[5] Z. El Mrabet, N. Kaabouch, H. El Ghazi, and H. El Ghazi, "Cyber-security in smart grid: Survey and challenges," *Computers & Electrical Engineering*, vol. 67, pp. 469–482, 2018.

[6] A. H. Abed, M. Nasr, and L. Abd Elhamid, "A conceptual framework for minimizing peak load electricity using Internet of Things," *International Journal of Computer Science and Mobile Computing*, vol. 10, no. 8, pp. 60–71, 2021.

[7] M. Meliani, A. E. Barkany, I. E. Abbassi, A. M. Darcherif, and M. Mahmoudi, "Energy management in the smart grid: State-of-the-art and future trends," *International Journal of Engineering Business Management*, vol. 13, p. 18479790211032920, 2021.

[8] S. K. Rathor and D. Saxena, "Energy management system for smart grid: An overview and key issues," *International Journal of Energy Research*, vol. 44, no. 6, pp. 4067–4109, 2020.

[9] D. Han, W. Sun, and X. Fan, "Dynamic energy management in smart grid: A fast randomized first-order optimization algorithm," *International Journal of Electrical Power & Energy Systems*, vol. 94, pp. 179–187, 2018.

[10]   I. Alotaibi, M. A. Abido, M. Khalid, and A. V. Savkin, "A comprehensive review of recent advances in smart grids: A sustainable future with renewable energy resources," *Energies*, vol. 13, no. 23, p. 6269, 2020.

[11]   J. S. Vardakas, N. Zorba, and C. V. Verikoukis, "A survey on demand response programs in smart grids: Pricing methods and optimization algorithms," *IEEE Communications Surveys & Tutorials*, vol. 17, no. 1, pp. 152–178, 2014.

[12]   A. Fakhar, A. M. Haidar, M. Abdullah, and N. Das, "Smart grid mechanism for green energy management: a comprehensive review," *International Journal of Green Energy*, vol. 20, no. 3, pp. 284–308, 2023.

[13]   S. A. Malik, T. M. Gondal, S. Ahmad, M. Adil, and R. Qureshi, "Towards optimization approaches in smart grid a review," in *2019 2nd International Conference on Computing, Mathematics and Engineering Technologies (iCoMET)*, 2019: IEEE, pp. 1–5.

[14]   H. J. Jabir, J. Teh, D. Ishak, and H. Abunima, "Impacts of demand-side management on electrical power systems: A review," *Energies*, vol. 11, no. 5, p. 1050, 2018.

[15]   A. R. Khan, A. Mahmood, A. Safdar, Z. A. Khan, S. Bilal, and N. A. K. Javaid, "Load forecasting and dynamic pricing based energy management in smart grid-a review," *in International Multi-topic Conference*, 2015.

[16]   G. S. Thirunavukkarasu, M. Seyedmahmoudian, E. Jamei, B. Horan, S. Mekhilef, and A. Stojcevski, "Role of optimization techniques in microgrid energy management systems—A review," *Energy Strategy Reviews*, vol. 43, p. 100899, 2022.

[17]   J. Powell, A. McCafferty-Leroux, W. Hilal, and S. A. Gadsden, "Smart grids: A comprehensive survey of challenges, industry applications, and future trends," *Energy Reports*, vol. 11, pp. 5760–5785, 2024.

[18]   S. F. Santos, D. Z. Fitiwi, M. Shafie-Khah, A. Bizuayehu, and J. Catalão, "Optimal sizing and placement of smart-grid-enabling technologies for maximizing renewable integration," in H.A. Gabbar (ed.), *Smart Energy Grid Engineering*. Elsevier, 2017, pp. 47–81.

[19]   Z. Ullah, R. Asghar, I. Khan, *et al.*, "Renewable energy resources penetration within smart grid: An overview," in *2020 International Conference on Electrical, Communication, and Computer Engineering (ICECCE)*, 2020: IEEE, pp. 1–6.

[20]   H. M. Hussain, N. Javaid, S. Iqbal, Q. U. Hasan, K. Aurangzeb, and M. Alhussein, "An efficient demand side management system with a new optimized home energy management controller in smart grid," *Energies*, vol. 11, no. 1, p. 190, 2018.

[21]   M. Awais, N. Javaid, K. Aurangzeb, S. I. Haider, Z. A. Khan, and D. Mahmood, "Towards effective and efficient energy management of single home and a smart community exploiting heuristic optimization algorithms with critical peak and real-time pricing tariffs in smart grids," *Energies*, vol. 11, no. 11, p. 3125, 2018.

[22]   F. Y. Melhem, "Optimization methods and energy management in smart grids," *Université Bourgogne Franche-Comté*, 2018.

[23] G. Venkataramani, P. Parankusam, V. Ramalingam, and J. Wang, "A review on compressed air energy storage–A pathway for smart grid and polygeneration," *Renewable and Sustainable Energy Reviews*, vol. 62, pp. 895–907, 2016.

[24] A. M. Eltamaly, "An accurate piecewise aging model for Li-ion batteries in hybrid renewable energy system applications," *Arabian Journal for Science and Engineering*, vol. 49, pp. 6551–6575, 2024.

[25] A. M. Eltamaly, "A novel energy storage and demand side management for entire green smart grid system for NEOM city in Saudi Arabia," *Energy Storage*, vol. 6, no. 1, p. e515, 2024.

[26] J. Every, L. Li, and D. G. Dorrell, "Leveraging smart meter data for economic optimization of residential photovoltaics under existing tariff structures and incentive schemes," *Applied Energy*, vol. 201, pp. 158–173, 2017.

[27] A. M. Eltamaly and M. A. Mohamed, "Optimal sizing and designing of hybrid renewable energy systems in smart grid applications," in I. Yahyaoui (ed.), *Advances in Renewable Energies and Power Technologies*. Elsevier, 2018, pp. 231–313.

[28] M. Abdelaziz Mohamed, A. M. Eltamaly, M. Abdelaziz Mohamed, and A. M. Eltamaly, "A PSO-based smart grid application for optimum sizing of hybrid renewable energy systems," in M. Abdelaziz Mohamed and A. M. Eltamaly (eds.), *Modeling and Simulation of Smart Grid Integrated with Hybrid Renewable Energy Systems*. Springer, 2018, pp. 53–60.

[29] A. M. Eltamaly, M. A. Alotaibi, A. I. Alolah, and M. A. Ahmed, "A novel demand response strategy for sizing of hybrid energy system with smart grid concepts," *IEEE Access*, vol. 9, pp. 20277–20294, 2021, doi:10.1109/Access.2021.3052128.

[30] A. M. Eltamaly and M. A. Alotaibi, "Novel fuzzy-swarm optimization for sizing of hybrid energy systems applying smart grid concepts," *IEEE Access*, vol. 9, pp. 93629–93650, 2021, doi:10.1109/Access.2021.3093169.

[31] A. M. Eltamaly, M. A. Alotaibi, W. A. Elsheikh, A. I. Alolah, and M. A. Ahmed, "Novel demand side-management strategy for smart grid concepts applications in hybrid renewable energy systems," *In 2022 4th International Youth Conference on Radio Electronics, Electrical and Power Engineering (REEPE)*. 2022: IEEE, pp. 1–7.

[32] J. Leitao, P. Gil, B. Ribeiro, and A. Cardoso, "A survey on home energy management," *IEEE Access*, vol. 8, pp. 5699–5722, 2020.

[33] N. H. Sonet, S. S. Rahman, A. Subhana, and F. Abid, "Prospect of demand side management in Bangladesh: A survey on residential application," in *2020 11th International Conference on Electrical and Computer Engineering (ICECE)*, 2020: IEEE, pp. 399–402.

[34] M. M. Iqbal, M. I. A. Sajjad, S. Amin, *et al.*, "Optimal scheduling of residential home appliances by considering energy storage and stochastically modelled photovoltaics in a grid exchange environment using hybrid grey wolf genetic algorithm optimizer," *Applied Sciences*, vol. 9, no. 23, p. 5226, 2019.

[35] M. Alizadeh, X. Li, Z. Wang, A. Scaglione, and R. Melton, "Demand-side management in the smart grid: Information processing for the power switch," *IEEE Signal Processing Magazine*, vol. 29, no. 5, pp. 55–67, 2012.

[36] P. Ravibabu, K. Venkatesh, T. Swetha, S. Kodad, and B. S. Ram, "Application of DSM techniques and renewable energy devises for peak load management," in *2008 IEEE Region 8 International Conference on Computational Technologies in Electrical and Electronics Engineering*, 2008: IEEE, pp. 129–132.

[37] M. Praveen and G. S. Rao, "Ensuring the reduction in peak load demands based on load shifting DSM strategy for smart grid applications," *Procedia Computer Science*, vol. 167, pp. 2599–2605, 2020.

[38] N. Tutkun, F. Ungören, and B. Alpagut, "Improved load shifting and valley filling strategies in demand side management in a nano scale off-grid wind-PV system in remote areas," in *2017 IEEE 14th International Conference on Networking, Sensing and Control (ICNSC)*, 2017: IEEE, pp. 13–18.

[39] M. H. Riaz, M. Zeeshan, T. Kamal, and S. A. H. Shah, "Demand side management using different energy conservation techniques," in *2017 International Multi-topic Conference (INMIC)*, 2017: IEEE, pp. 1–4.

[40] H. Attia, M. El-Sobki, M. El-Metwally, and S. Wahdan, "Tariff design for load building/energy conservation a DSM formulation," in *2008 12th International Middle-East Power System Conference*, 2008: IEEE, pp. 571–576.

[41] P. Shanmugapriya, M. S. Kumaran, J. Baskaran, C. Nayanatara, P. Sharmila, and A. M. Eltamaly, "Flexible dispatch strategy adopted by optimizing DG parameters in a real time power system distributed network," *Journal of Electrical Engineering & Technology*, vol. 17, no. 2, pp. 847–861, 2022, doi:10.1007/s42835–021–00938–8.

[42] M. A. Alotaibi and A. M. Eltamaly, "A smart strategy for sizing of hybrid renewable energy system to supply remote loads in Saudi Arabia," *Energies*, vol. 14, no. 21, 2021, doi:10.3390/en14217069.

[43] Q. Qdr, "Benefits of demand response in electricity markets and recommendations for achieving them," *U.S. Department of Energy*, Washington, DC, USA, Tech. Rep, vol. 2006, p. 95, 2006.

[44] U. Assad, M. A. S. Hassan, U. Farooq, *et al.*, "Smart grid, demand response and optimization: a critical review of computational methods," *Energies*, vol. 15, no. 6, p. 2003, 2022.

[45] A. Malik and J. Ravishankar, "A review of demand response techniques in smart grids," in *2016 IEEE Electrical Power and Energy Conference (EPEC)*, 2016: IEEE, pp. 1–6.

[46] A. M. Eltamaly and A. N. A. Elghaffar, "Basic definitions of smart grid technologies and applications." *2nd International Baku Conference on Scientific Research*, 2021.

[47] S. Datchanamoorthy, S. Kumar, Y. Ozturk, and G. Lee, "Optimal time-of-use pricing for residential load control," in *2011 IEEE International Conference on Smart Grid Communications (SmartGridComm)*, 2011: IEEE, pp. 375–380.

[48] P. K. Wesseh Jr and B. Lin, "A time-of-use pricing model of the electricity market considering system flexibility," *Energy Reports*, vol. 8, pp. 1457–1470, 2022.

[49] L. I. Hagfors, H. H. Kamperud, F. Paraschiv, M. Prokopczuk, A. Sator, and S. Westgaard, "Prediction of extreme price occurrences in the German day-ahead electricity market," *Quantitative Finance*, vol. 16, no. 12, pp. 1929–1948, 2016.

[50] J. Wen, X.-X. Zhao, and C.-P. Chang, "The impact of extreme events on energy price risk," *Energy Economics*, vol. 99, p. 105308, 2021.

[51] C. Joe-Wong, S. Sen, S. Ha, and M. Chiang, "Optimized day-ahead pricing for smart grids with device-specific scheduling flexibility," *IEEE Journal on Selected Areas in Communications*, vol. 30, no. 6, pp. 1075–1085, 2012.

[52] T.-C. Chiu, Y.-Y. Shih, A.-C. Pang, and C.-W. Pai, "Optimized day-ahead pricing with renewable energy demand-side management for smart grids," *IEEE Internet of Things Journal*, vol. 4, no. 2, pp. 374–383, 2016.

[53] S. Caron and G. Kesidis, "Incentive-based energy consumption scheduling algorithms for the smart grid," in *2010 First IEEE International Conference on Smart Grid Communications*, 2010: IEEE, pp. 391–396.

[54] C. Chen, J. Wang, and S. Kishore, "A distributed direct load control approach for large-scale residential demand response," *IEEE Transactions on Power Systems*, vol. 29, no. 5, pp. 2219–2228, 2014.

[55] A. Haque, M. Nijhuis, G. Ye, P. H. Nguyen, F. W. Bliek, and J. G. Slootweg, "Integrating direct and indirect load control for congestion management in LV networks," *IEEE Transactions on Smart Grid*, vol. 10, no. 1, pp. 741–751, 2017.

[56] J. Aghaei, M.-I. Alizadeh, P. Siano, and A. Heidari, "Contribution of emergency demand response programs in power system reliability," *Energy*, vol. 103, pp. 688–696, 2016.

[57] X. Lu, X. Ge, K. Li, *et al.*, "Optimal bidding strategy of demand response aggregator based on customers' responsiveness behaviors modeling under different incentives," *IEEE Transactions on Industry Applications*, vol. 57, no. 4, pp. 3329–3340, 2021.

[58] R. Hornby, C. James, K. Takahashi, and D. White, "Increasing demand response in Maine," 2008. Available from: https://www.synapse-energy.com/sites/default/files/SynapseReport.2008–01.ME-PUC.Increasing-Demand-Response-in-Maine.07–077.pdf.

[59] T. Cortes-Arcos, J. L. Bernal-Agustín, R. Dufo-López, J. M. Lujano-Rojas, and J. Contreras, "Multi-objective demand response to real-time prices (RTP) using a task scheduling methodology," *Energy*, vol. 138, pp. 19–31, 2017.

[60] S. Lu, N. Samaan, R. Diao, *et al.*, "Centralized and decentralized control for demand response," *in ISGT 2011*, 2011: IEEE, pp. 1–8.

[61] L. C. Siebert, L. R. Ferreira, E. K. Yamakawa, *et al.*, "Centralized and decentralized approaches to demand response using smart plugs," in *2014 IEEE PES T&D Conference and Exposition*, 2014: IEEE, pp. 1–5.

[62]   N. Javaid, G. Hafeez, S. Iqbal, N. Alrajeh, M. S. Alabed, and M. Guizani, "Energy efficient integration of renewable energy sources in the smart grid for demand side management," *IEEE Access*, vol. 6, pp. 77077–77096, 2018.

[63]   C. W. Gellings, "The concept of demand-side management for electric utilities," *Proceedings of the IEEE*, vol. 73, no. 10, pp. 1468–1470, 1985.

[64]   Q. Liu, L. Ma, L. Li, and Y. Liu, "Optimal robust real-time pricing algorithm based on utility maximization for smart grid," in *2021 4th International Conference on Energy, Electrical and Power Engineering (CEEPE)*, 2021: IEEE, pp. 836–841.

[65]   A.-H. Mohsenian-Rad, V. W. Wong, J. Jatskevich, and R. Schober, "Optimal and autonomous incentive-based energy consumption scheduling algorithm for smart grid," in *2010 Innovative Smart Grid Technologies (ISGT)*, 2010: IEEE, pp. 1–6.

[66]   S. Esmer and E. Bekiroglu, "Design of PMaSynRM for flywheel energy storage system in smart grids," *International Journal of Smart Grid*, vol. 6, no. 4, pp. 84–91, 2022.

[67]   F. Tooryan, H. HassanzadehFard, E. R. Collins, S. Jin, and B. Ramezani, "Smart integration of renewable energy resources, electrical, and thermal energy storage in microgrid applications," *Energy*, vol. 212, p. 118716, 2020.

[68]   A. Lyden, C. Brown, I. Kolo, G. Falcone, and D. Friedrich, "Seasonal thermal energy storage in smart energy systems: District-level applications and modelling approaches," *Renewable and Sustainable Energy Reviews*, vol. 167, p. 112760, 2022.

[69]   G. Navarro, J. Torres, M. Blanco, J. Nájera, M. Santos-Herran, and M. Lafoz, "Present and future of supercapacitor technology applied to powertrains, renewable generation and grid connection applications," *Energies*, vol. 14, no. 11, p. 3060, 2021.

[70]   H. Chamandoust, A. Hashemi, and S. Bahramara, "Energy management of a smart autonomous electrical grid with a hydrogen storage system," *International Journal of Hydrogen Energy*, vol. 46, no. 34, pp. 17608–17626, 2021.

[71]   M. Rezaeimozafar, R. F. Monaghan, E. Barrett, and M. Duffy, "A review of behind-the-meter energy storage systems in smart grids," *Renewable and Sustainable Energy Reviews*, vol. 164, p. 112573, 2022.

[72]   B. P. Roberts and C. Sandberg, "The role of energy storage in development of smart grids," *Proceedings of the IEEE*, vol. 99, no. 6, pp. 1139–1144, 2011.

[73]   M. G. Molina, "Distributed energy storage systems for applications in future smart grids," in *2012 Sixth IEEE/PES Transmission and Distribution: Latin America Conference and Exposition (T&D-LA)*, 2012: IEEE, pp. 1–7.

[74]   E. Ozdemir, S. Ozdemir, K. Erhan, and A. Aktas, "Energy storage technologies opportunities and challenges in smart grids," *in 2016 International Smart Grid Workshop and Certificate Program (ISGWCP)*, 2016: IEEE, pp. 1–6.

[75]   M. Abdelaziz Mohamed, A. M. Eltamaly, M. Abdelaziz Mohamed, and A. M. Eltamaly, "A novel smart grid application for optimal sizing of hybrid renewable energy systems," in M. Abdelaziz Mohamed and

A. M. Eltamaly (eds.), *Modeling and Simulation of Smart Grid Integrated with Hybrid Renewable Energy Systems.* Springer, 2018, pp. 39–51.

[76] A. M. Eltamaly, "Smart decentralized electric vehicle aggregators for optimal dispatch technologies," *Energies*, vol. 16, no. 24, p. 8112, 2023.

[77] A. M. Eltamaly, "Optimal dispatch strategy for electric vehicles in V2G applications," *Smart Cities*, vol. 6, no. 6, pp. 3161–3191, 2023.

[78] Z. A. Almutairi, A. M. Eltamaly, A. El Khereiji, *et al.*, "Modeling and experimental determination of lithium-ion battery degradation in hot environment," In *2022 23rd International Middle East Power Systems Conference (MEPCON)*, 2022: IEEE, pp. 1–8.

[79] A. Maheri, "Multi-objective design optimisation of standalone hybrid wind-PV-diesel systems under uncertainties," *Renewable Energy*, vol. 66, pp. 650–661, 2014.

[80] A. Kamjoo, A. Maheri, A. M. Dizqah, and G. A. Putrus, "Multi-objective design under uncertainties of hybrid renewable energy system using NSGA-II and chance constrained programming," *International Journal of Electrical Power & Energy Systems*, vol. 74, pp. 187–194, 2016.

[81] A. Maleki, M. G. Khajeh, and M. Ameri, "Optimal sizing of a grid independent hybrid renewable energy system incorporating resource uncertainty, and load uncertainty," *International Journal of Electrical Power & Energy Systems*, vol. 83, pp. 514–524, 2016.

[82] A. Arabali, M. Ghofrani, M. Etezadi-Amoli, and M. S. Fadali, "Stochastic performance assessment and sizing for a hybrid power system of solar/wind/energy storage," *IEEE Transactions on Sustainable Energy*, vol. 5, no. 2, pp. 363–371, 2013.

[83] K.-H. Chang and G. Lin, "Optimal design of hybrid renewable energy systems using simulation optimization," *Simulation Modelling Practice and Theory*, vol. 52, pp. 40–51, 2015.

[84] H. El-Tamaly, A. M. El-Tamaly, and A. El-Baset Mohammed, "Design and control strategy of utility interfaced PV/WTG hybrid system," in *The Ninth International Middle East Power System Conference, MEPCON*, 2003, pp. 16–18.

[85] A. M. Eltamaly and M. Mohamed, "A novel software for design and optimization of hybrid power systems," *Journal of the Brazilian Society of Mechanical Sciences and Engineering*, vol. 38, no. 4, pp. 1299–1315, 2016, doi:10.1007/s40430–015–0363-z.

[86] A. M. Eltamaly and A. A. Al-Shamma'a, "Optimal configuration for isolated hybrid renewable energy systems," *Journal of Renewable and Sustainable Energy*, vol. 8, no. 4, p. 045502, 2016.

[87] A. M. Eltamaly, K. E. Addoweesh, U. Bawa, and M. A. Mohamed, "Economic modeling of hybrid renewable energy system: A case study in Saudi Arabia," *Arabian Journal for Science and Engineering*, vol. 39, no. 5, pp. 3827–3839, 2014, doi:10.1007/s13369–014–0945-6.

[88] A. M. Eltamaly, "Design and implementation of wind energy system in Saudi Arabia," *Renewable Energy*, vol. 60, pp. 42–52, 2013, doi:10.1016/j.renene.2013.04.006.

[89]   U. Khaled, A. M. Eltamaly, and A. Beroual, "Optimal power flow using particle swarm optimization of renewable hybrid distributed generation," *Energies*, vol. 10, no. 7, 2017, doi:10.3390/en10071013.

[90]   M. A. Mohamed, A. M. Eltamaly, and A. I. Alolah, "PSO-based smart grid application for sizing and optimization of hybrid renewable energy systems," *PLoS One*, vol. 11, no. 8, p. e0159702, 2016, doi:10.1371/journal. pone.0159702.

[91]   A. M. Eltamaly and M. A. Mohamed, Alolah, A.I., "A novel smart grid theory for optimal sizing of hybrid renewable energy systems," *Solar Energy*, vol. 124, pp. 26–38, 2016.

[92]   H. A. El-Sattar, S. Kamel, H. Sultan, M. Tostado-Véliz, A. M. Eltamaly, and F. Jurado, "Performance analysis of a stand-alone PV/WT/Biomass/Bat system in Alrashda Village in Egypt," *Applied Sciences*, vol. 11, no. 21, p. 10191, 2021, doi:10.3390/app112110191.

[93]   B. Naghibi, M. A. Masoum, and S. Deilami, "Effects of V2H integration on optimal sizing of renewable resources in smart home based on Monte Carlo simulations," *IEEE Power and Energy Technology Systems Journal*, vol. 5, no. 3, pp. 73–84, 2018.

[94]   S. S. Singh and E. Fernandez, "Modeling, size optimization and sensitivity analysis of a remote hybrid renewable energy system," *Energy*, vol. 143, pp. 719–731, 2018.

[95]   L. Urbanucci and D. Testi, "Optimal integrated sizing and operation of a CHP system with Monte Carlo risk analysis for long-term uncertainty in energy demands," *Energy Conversion and Management*, vol. 157, pp. 307–316, 2018.

[96]   T.-C. Chen, M. Ramesh, C. Qin, *et al.*, "Smart sampling of representative hourly power generation scenario with high renewable penetration," *IEEE Access*, 2024.

[97]   J. Cai, L. Hao, Q. Xu, and K. Zhang, "Reliability assessment of renewable energy integrated power systems with an extendable Latin hypercube importance sampling method," *Sustainable Energy Technologies and Assessments*, vol. 50, p. 101792, 2022.

[98]   K. Kappner, P. Letmathe, and P. Weidinger, "Optimisation of photovoltaic and battery systems from the prosumer-oriented total cost of ownership perspective," *Energy, Sustainability and Society*, vol. 9, pp. 1–24, 2019.

[99]   A. M. Eltamaly, E. Ali, M. Bumazza, S. Mulyono, and M. Yasin, "Optimal design of hybrid renewable energy system for a reverse osmosis desalination system in Arar, Saudi Arabia," *Arabian Journal for Science and Engineering*, vol. 46, no. 10, pp. 9879–9897, 2021, doi:10.1007/s13369-021-05645-0.

[100] A. M. Eltamaly, A. Y. Abdelaziz, and A. G. Abo-Khalil, *Control and Operation of Grid-Connected Wind Energy Systems*. Cham: Springer, 2021.

[101] A. M. Eltamaly, K. E. Addoweesh, U. Bawah, and M. A. Mohamed, "New software for hybrid renewable energy assessment for ten locations in Saudi Arabia," *Journal of Renewable and Sustainable Energy*, vol. 5, no. 3, p. 033126, 2013, doi:10.1063/1.4809791.

[102]   M. A. Alotaibi and A. M. Eltamaly, "Upgrading conventional power system for accommodating electric vehicle through demand side management and V2G concepts," *Energies*, vol. 15, no. 18, 2022, doi:10.3390/en15186541.

[103]   G. Tina, S. Gagliano, and S. Raiti, "Hybrid solar/wind power system probabilistic modelling for long-term performance assessment," *Solar Energy*, vol. 80, no. 5, pp. 578–588, 2006.

[104]   L. Wang and C. Singh, "Multicriteria design of hybrid power generation systems based on a modified particle swarm optimization algorithm," *IEEE Transactions on Energy Conversion*, vol. 24, no. 1, pp. 163–172, 2009.

[105]   H. Yang, W. Zhou, L. Lu, and Z. Fang, "Optimal sizing method for stand-alone hybrid solar–wind system with LPSP technology by using genetic algorithm," *Solar Energy*, vol. 82, no. 4, pp. 354–367, 2008.

[106]   S. Karaki, R. Chedid, and R. Ramadan, "Probabilistic performance assessment of autonomous solar-wind energy conversion systems," *IEEE Transactions on Energy Conversion*, vol. 14, no. 3, pp. 766–772, 1999.

[107]   L. N. Kishore and E. Fernandez, "Reliability well-being assessment of PV-wind hybrid system using Monte Carlo simulation," in *2011 International Conference on Emerging Trends in Electrical and Computer Technology*, 2011: IEEE, pp. 63–68.

[108]   H. H. El-Tamaly, M. Hamada, and A. M. ELtamaly, "Computer simulation of wind energy system and applications," in *Proceedings of the International AMSE Conference on System Analysis, Control & Designs (AMSE)*, vol. 4, pp. 84–94, 1995.

[109]   R. Chedid and S. Rahman, "Unit sizing and control of hybrid wind-solar power systems," *IEEE Transactions on Energy Conversion*, vol. 12, no. 1, pp. 79–85, 1997.

[110]   F. Huneke, J. Henkel, J. A. Benavides González, and G. Erdmann, "Optimisation of hybrid off-grid energy systems by linear programming," *Energy, Sustainability and Society*, vol. 2, pp. 1–19, 2012.

[111]   P. Kumar and S. Deokar, "Designing and simulation tools of renewable energy systems: review literature," *Progress in Advanced Computing and Intelligent Engineering: Proceedings of ICACIE 2016*, vol. 1, pp. 315–324, 2018.

[112]   F. A. Khan, N. Pal, and S. H. Saeed, "Review of solar photovoltaic and wind hybrid energy systems for sizing strategies optimization techniques and cost analysis methodologies," *Renewable and Sustainable Energy Reviews*, vol. 92, pp. 937–947, 2018.

[113]   M. Mudasiru, "Monte-Carlo based robust analytical method for optimal sizing and reliability of hybrid renewable energy system," Universiti Teknologi *Malaysia*, 2020.

[114]   Z. Zhao, N. Holland, and J. Nelson, "Optimizing smart grid performance: A stochastic approach to renewable energy integration," *Sustainable Cities and Society*, p. 105533, 2024.

[115]   A. Keyvandarian and A. Saif, "An adaptive distributionally robust optimization approach for optimal sizing of hybrid renewable energy systems,"

Journal of Optimization Theory and Applications, vol. 203, pp. 2055–2082, 2024.

[116] H. Shuai, J. Fang, X. Ai, Y. Tang, J. Wen, and H. He, "Stochastic optimization of economic dispatch for microgrid based on approximate dynamic programming," *IEEE Transactions on Smart Grid*, vol. 10, no. 3, pp. 2440–2452, 2018.

[117] A. Cervone, G. Carbone, E. Santini, and S. Teodori, "Optimization of the battery size for PV systems under regulatory rules using a Markov-Chains approach," *Renewable Energy*, vol. 85, pp. 657–665, 2016.

[118] J. Dong, F. Gao, X. Guan, Q. Zhai, and J. Wu, "Storage sizing with peak-shaving policy for wind farm based on cyclic Markov chain model," *IEEE Transactions on Sustainable Energy*, vol. 8, no. 3, pp. 978–989, 2016.

[119] J. Tomić and W. Kempton, "Using fleets of electric-drive vehicles for grid support," *Journal of Power Sources*, vol. 168, no. 2, pp. 459–468, 2007.

[120] H. K. Nunna, S. Battula, S. Doolla, and D. Srinivasan, "Energy management in smart distribution systems with vehicle-to-grid integrated microgrids," *IEEE Transactions on Smart Grid*, vol. 9, no. 5, pp. 4004–4016, 2016.

[121] R. Moreira, L. Ollagnier, D. Papadaskalopoulos, and G. Strbac, "Optimal multi-service business models for electric vehicles," In *2017 IEEE Manchester PowerTech*, 2017: IEEE, pp. 1–6.

[122] A. Y. Saber and G. K. Venayagamoorthy, "Intelligent unit commitment with vehicle-to-grid—A cost-emission optimization," *Journal of Power Sources*, vol. 195, no. 3, pp. 898–911, 2010.

[123] M. A. López, S. De La Torre, S. Martín, and J. A. Aguado, "Demand-side management in smart grid operation considering electric vehicles load shifting and vehicle-to-grid support," *International Journal of Electrical Power & Energy Systems*, vol. 64, pp. 689–698, 2015.

[124] L. Calearo, A. Thingvad, H. H. Ipsen, and M. Marinelli, "Economic value and user remuneration for EV based distribution grid services," *In 2019 IEEE PES Innovative Smart Grid Technologies Europe (ISGT-Europe)*, 2019: IEEE, pp. 1–5.

[125] M. A. Ortega-Vazquez, "Optimal scheduling of electric vehicle charging and vehicle-to-grid services at household level including battery degradation and price uncertainty," *IET Generation, Transmission & Distribution*, vol. 8, no. 6, pp. 1007–1016, 2014.

[126] K. Young, C. Wang, L. Y. Wang, and K. Strunz, "Electric vehicle battery technologies," in R. Garcia-Valle, J. A. Peças Lopes (eds.), *Electric Vehicle Integration into Modern Power Networks*. Springer, 2012, pp. 15–56.

[127] M. Yilmaz and P. T. Krein, "Review of the impact of vehicle-to-grid technologies on distribution systems and utility interfaces," *IEEE Transactions on Power Electronics*, vol. 28, no. 12, pp. 5673–5689, 2012.

[128] N. Hartmann and E. D. Özdemir, "Impact of different utilization scenarios of electric vehicles on the German grid in 2030," *Journal of Power Sources*, vol. 196, no. 4, pp. 2311–2318, 2011.

[129] C. Weiller, "Plug-in hybrid electric vehicle impacts on hourly electricity demand in the United States," *Energy Policy*, vol. 39, pp. 3766–3778, 2011.

[130] L. Zhu, J. He, L. He, W. Huang, Y. Wang, and Z. Liu, "Optimal operation strategy of PV-charging-hydrogenation composite energy station considering demand response," *Energies*, vol. 15, no. 16, p. 5915, 2022.

[131] I. Karakitsios, D. Lagos, A. Dimeas, and N. Hatziargyriou, "How can EVs support high RES penetration in islands," *Energies*, vol. 16, no. 1, p. 558, 2023.

[132] M. S. Alam, F. S. Al-Ismail, A. Salem, and M. A. Abido, "High-level penetration of renewable energy sources into grid utility: Challenges and solutions," *IEEE Access*, vol. 8, pp. 190277–190299, 2020.

[133] J. L. Bernal-Agustín and R. Dufo-Lopez, "Simulation and optimization of stand-alone hybrid renewable energy systems," *Renewable and Sustainable Energy Reviews*, vol. 13, no. 8, pp. 2111–2118, 2009.

[134] S. Sinha and S. Chandel, "Review of software tools for hybrid renewable energy systems," *Renewable and Sustainable Energy Reviews*, vol. 32, pp. 192–205, 2014.

[135] D. Kaur and P. Cheema, "Software tools for analyzing the hybrid renewable energy sources:-a review," in *2017 International Conference on Inventive Systems and Control (ICISC)*, 2017: IEEE, pp. 1–4.

[136] A. Mahesh and K. S. Sandhu, "Hybrid wind/photovoltaic energy system developments: Critical review and findings," *Renewable and Sustainable Energy Reviews*, vol. 52, pp. 1135–1147, 2015.

[137] A. A. Khan, A. F. Minai, R. K. Pachauri, and H. Malik, "Optimal sizing, control, and management strategies for hybrid renewable energy systems: A comprehensive review," *Energies*, vol. 15, no. 17, p. 6249, 2022.

[138] H. Shahinzadeh, M. Moazzami, S. H. Fathi, and G. B. Gharehpetian, "Optimal sizing and energy management of a grid-connected microgrid using HOMER software," in *2016 Smart Grids Conference (SGC)*, 2016: IEEE, pp. 1–6.

[139] O. Boqtob, H. El Moussaoui, H. El Markhi, and T. Lamhamdi, "Optimal sizing of grid connected microgrid in Morocco using Homer Pro," in *2019 international conference on wireless technologies, embedded and intelligent systems (WITS)*, 2019: IEEE, pp. 1–6.

[140] M. G. Yenalem, M. Ngoo, D. Shiferaw, and P. Hinga, "Modelling and optimal sizing of grid-connected micro grid system using HOMER in Bahir Dar City, Ethiopia," *International Journal of Power Systems*, vol. 5, pp. 1–12, 2020.

[141] I. C. Hoarcă, N. Bizon, I. S. Șorlei, and P. Thounthong, "Sizing design for a hybrid renewable power system using HOMER and iHOGA simulators," *Energies*, vol. 16, no. 4, p. 1926, 2023.

[142] H. Shamachurn, "Optimization of an off-grid domestic Hybrid Energy System in suburban Paris using iHOGA software," *Renewable Energy Focus*, vol. 37, pp. 36–49, 2021.

[143] P. Ganguly, A. Kalam, and A. Zayegh, "Design an optimum standalone hybrid renewable energy system for a small town at Portland, Victoria using iHOGA," in *2017 Australasian Universities Power Engineering Conference (AUPEC)*, 2017: IEEE, pp. 1–6.

[144] A. Mills and S. Al-Hallaj, "Simulation of hydrogen-based hybrid systems using Hybrid2," *International Journal of Hydrogen Energy*, vol. 29, no. 10, pp. 991–999, 2004.

[145] H. Dinçer, S. Yüksel, T. Aksoy, Ü. Hacıoğlu, A. Mikhaylov, and G. Pinter, "Analysis of solar module alternatives for efficiency-based energy investments with hybrid 2-tuple IVIF modeling," *Energy Reports*, vol. 10, pp. 61–71, 2023.

[146] F. Zaro and N. A. Ayyash, "Design and management of hybrid renewable energy system using RETscreen software: A case study," *International Journal of Electrical Engineering and Computer Science*, vol. 5, pp. 164–170, 2023.

[147] S. U.-D. Khan, I. Wazeer, and Z. Almutairi, "Comparative analysis of SAM and RETScreen tools for the case study of 600 kW solar PV system installation in Riyadh, Saudi Arabia," *Sustainability*, vol. 15, no. 6, p. 5381, 2023.

[148] A. Boussaibo and A. D. Pene, "Optimal sizing and power losses reduction of photovoltaic systems using PVsyst software," *Journal of Power and Energy Engineering*, vol. 12, no. 7, pp. 23–38, 2024.

[149] M. Baqir and H. K. Channi, "Analysis and design of solar PV system using PVsyst software," *Materials Today: Proceedings*, vol. 48, pp. 1332–1338, 2022.

[150] S. Baviskar, G. B. Patil, Y. Patil, Y. Solanki, and S. Ushkewar, "An evaluation of solar photovoltaic system depreciation using PVSOL," *International Journal of Engineering and Management Research*, vol. 14, no. 3, pp. 15–20, 2024.

[151] M. M. Ibrahim, "Investigation of a grid-connected solar PV system for the electric-vehicle charging station of an office building using PVSOL software," *Polityka Energetyczna (Energy Policy Journal)*, vol. 25, no. 1, pp. 175–208, 2022.

[152] J. Singhal, K. Barva, and S. Joshi, "Hybrid renewable energy systems through SAM software: A comprehensive approach," in *2024 IEEE International Conference on Electronics, Computing and Communication Technologies (CONECCT)*, 2024: IEEE, pp. 1–6.

[153] C. Tíba, A. L. B. Candeias, N. Fraidenraich, E. d. S. Barbosa, P. B. de Carvalho Neto, and J. B. de Melo Filho, "A GIS-based decision support tool for renewable energy management and planning in semi-arid rural environments of northeast of Brazil," *Renewable Energy*, vol. 35, no. 12, pp. 2921–2932, 2010.

[154] R. G. S. Rao, N. Rayaguru, N. Renganathan, and S. Thakur, "City—scale spatial data infrastructure for solar photovoltaic energy generation assessment," *International Journal of Engineering and Technology*, vol. 7, pp. 4–7, 2018.

[155] A. Rana and G. Gróf, "Assessment of prosumer-based energy system for rural areas by using TRNSYS software," *Cleaner Energy Systems*, vol. 8, p. 100110, 2024.

[156] M. S. Saleem, N. Abas, A. R. Kalair, *et al.*, "Design and optimization of hybrid solar-hydrogen generation system using TRNSYS," *International Journal of Hydrogen Energy*, vol. 45, no. 32, pp. 15814–15830, 2020.

[157]   J. Torres, A. A. Recalde, and I. Endara, "Optimal sizing of distributed photovoltaic generation in a mv network," in *2020 IEEE PES Transmission & Distribution Conference and Exhibition-Latin America (T&D LA)*, 2020: IEEE, pp. 1–6.

[158]   A. Jain, A. Mani, and A. S. Siddiqui, "Simulation of a microgrid with OpenDSS an open-source software package," in R. Agrawal, C. Kishore Singh, A. Goyal, and D. K. Singh (eds), *Modern Electronics Devices and Communication Systems: Select Proceedings of MEDCOM 2021*. Springer, 2023, pp. 513–529.

[159]   V. A. H. Vajjala and W. Jewell, "Demand response potential in aggregated residential houses using GridLAB-D," in *2015 IEEE Conference on Technologies for Sustainability (SusTech)*, 2015: IEEE, pp. 27–34.

[160]   M. Jdeed, E. Sharma, C. Klemenjak, and W. Elmenreich, "Smart grid modeling and simulation—Comparing GridLAB-D and RAPSim via two Case studies," in *2018 IEEE International Energy Conference (ENERGY-CON)*, 2018: IEEE, pp. 1–6.

[161]   V. Pavlovsky, A. Prykhodko, O. Lenga, and A. Zakharov, "Automation approach of RES hosting capacity estimation in the distribution network using PowerFactory software," in *CIRED 2024 Vienna Workshop*, 2024: IEEE, pp. 569–572.

[162]   W. H. Tee, K. A. Qaid, C. K. Gan, and P. H. Tan, "Battery energy storage system sizing using PSO algorithm in DIgSILENT PowerFactory," *International Journal of Renewable Energy Research*, vol. 12, no. 4, pp. 2142–2150, 2022.

[163]   F. E. Abrahamsen, Y. Ai, and M. Cheffena, "Communication technologies for smart grid: A comprehensive survey," *Sensors*, vol. 21, no. 23, p. 8087, 2021.

[164]   Q.-D. Ho, Y. Gao, and T. Le-Ngoc, "Challenges and research opportunities in wireless communication networks for smart grid," *IEEE Wireless Communications*, vol. 20, no. 3, pp. 89–95, 2013.

[165]   M. Kuzlu, M. Pipattanasomporn, and S. Rahman, "Communication network requirements for major smart grid applications in HAN, NAN and WAN," *Computer Networks*, vol. 67, pp. 74–88, 2014.

[166]   R. H. Khan and J. Y. Khan, "A comprehensive review of the application characteristics and traffic requirements of a smart grid communications network," *Computer Networks*, vol. 57, no. 3, pp. 825–845, 2013.

[167]   F. Bouhafs, M. Mackay, and M. Merabti, "Links to the future: Communication requirements and challenges in the smart grid," *IEEE Power and Energy Magazine*, vol. 10, no. 1, pp. 24–32, 2011.

[168]   Y. Saleem, N. Crespi, M. H. Rehmani, and R. Copeland, "Internet of things-aided smart grid: technologies, architectures, applications, prototypes, and future research directions," *IEEE Access*, vol. 7, pp. 62962–63003, 2019.

[169]   A. M. Eltamaly and M. A. Ahmed, "Performance evaluation of communication infrastructure for peer-to-peer energy trading in community microgrids," *Energies*, vol. 16, no. 13, 2023, doi:10.3390/en16135116.

[170] F. Condon, J. M. Martínez, A. M. Eltamaly, Y. C. Kim, and M. A. Ahmed, "Design and implementation of a cloud-IoT-based home energy management system," *Sensors*, vol. 23, no. 1, 2023, doi:10.3390/s23010176.

[171] F. Condon, P. Franco, J. M. Martínez, A. M. Eltamaly, Y. C. Kim, and M. A. Ahmed, "EnergyAuction: IoT-blockchain architecture for local peer-to-peer energy trading in a microgrid," *Sustainability*, vol. 15, no. 17, 2023, doi:10.3390/su151713203.

[172] M. A. Ahmed, S. A. Chavez, A. M. Eltamaly, H. O. Garces, A. J. Rojas, and Y. C. Kim, "Toward an intelligent campus: IoT platform for remote monitoring and control of smart buildings," *Sensors*, vol. 22, no. 23, 2022, doi:10.3390/s22239045.

[173] A. M. Eltamaly, M. A. Alotaibi, A. I. Alolah, and M. A. Ahmed, "IoT-based hybrid renewable energy system for smart campus," *Sustainability*, vol. 13, no. 15, 2021, doi:10.3390/su13158555.

[174] A. M. Eltamaly, M. A. Ahmed, M. A. Alotaibi, A. Alolah, and Y. C. Kim, "Performance of communication network for monitoring utility scale photovoltaic power plants," *Energies*, vol. 13, no. 21, 2020, doi:10.3390/en13215527.

[175] M. A. Ahmed, A. M. Eltamaly, M. A. Alotaibi, A. I. Alolah, and Y. C. Kim, "Wireless network architecture for cyber physical wind energy system," *IEEE Access*, vol. 8, pp. 40180–40197, 2020, doi:10.1109/Access.2020.2976742.

Chapter 2

Optimization algorithms for optimal sizing of smart grid considering demand-side management

Ahmed A. Zaki Diab[1,2], Ali M. Eltamaly[3,4], Sultan I. EL-Ajmi[1] and Ahmed Khaled[5]

Abstract

This chapter investigates the impact of demand-side management (DSM) on the optimization of hybrid renewable energy systems. DSM is presented as a strategic approach to optimize energy consumption by shifting demand in accordance with renewable energy generation patterns, thereby improving system efficiency and reducing reliance on fossil fuels. The methods of DSM are detailed, and the Salp swarm algorithm (SSA) is applied to determine the optimal configuration of the energy system, integrating renewable energy sources, energy storage, and conventional generators. A real case study in New Minia, Egypt, is used to demonstrate the practical application of DSM in a hybrid energy system. The results show that DSM leads to a reduction in the cost of energy and diesel generator (DG) operating costs while improving energy utilization efficiency by eliminating the dummy load ratio. Despite a slight increase in the loss of power supply probability (LPSP), DSM maintains system reliability and provides significant economic and environmental benefits. The increase in the LPSP is caused because the system has been optimally designed using the real load data before applying the DSM. The findings highlight DSM as a cost-effective and sustainable strategy for optimizing hybrid energy systems in regions with growing renewable energy penetration. The chapter also includes a number of useful MATLAB® codes that are essential for the implementation of the system, providing practical tools for simulating and optimizing the hybrid renewable energy system with DSM. These codes enable efficient modeling, analysis, and integration of various system components, facilitating the application of DSM strategies in real-world scenarios.

[1]Electrical Engineering Department, Minia University, Egypt
[2]Department of Mechatronics Engineering, Minia National University, Egypt
[3]Electrical Engineering Department, Mansoura University, Egypt
[4]Sustainable Energy Technology Center, King Saud University, Saudi Arabia
[5]Department of Electrical Engineering, Faculty of Engineering, Al-Azhar University, Egypt

Keywords: DSM; COE; PV; Wind; Microgrid; Storage; Battery; MATLAB codes; Renewable energy sources; Optimization; Energy

2.1 Introduction

To foster the development of sustainable and attractive environments powered by renewable energy (RE) sources, many countries have launched initiatives to expand remote regions and establish new residential cities to accommodate population growth [1]. A range of feasibility studies and strategic plans have been conducted to explore the possibility of electrifying these remote areas by integrating them with the national grid. However, these studies often reveal that connecting to the grid can be prohibitively costly and may face technical challenges [1,2]. Additionally, the continued dependence on fossil fuels—such as diesel, natural gas, coal, mazut, and oil—for electricity generation, transportation, and industrial processes [3] significantly exacerbates climate change. These finite energy sources not only pollute the environment through greenhouse gas emissions, particularly carbon dioxide, but also contribute to long-term environmental degradation. As a result, many countries are increasingly turning to natural, clean, reliable, and cost-free energy alternatives. These include bioenergy, heat pumping technologies, geothermal energy, tidal energy, green hydrogen, hydropower, ocean energy, photovoltaic (PV) systems, solar heating and cooling, concentrated solar power, and wind energy. These renewable energy solutions offer promising pathways toward sustainable development and reducing reliance on fossil [4,5].

The International Energy Agency has forecasted substantial growth in renewable energy, with an estimated 3,700 GW of additional renewable capacity to be installed globally between 2023 and 2028. This marks a significant transition towards sustainable energy, with projections indicating that over 42% of the world's energy will be derived from renewable sources by 2028. Solar PV and wind energy, in particular, are expected to grow at annual rates of approximately 9.5% and 6.1%, respectively, from 2020 to 2028 [2,6]. These two renewable energy sources are key components of Egypt's energy strategy, as the country aims to expand its solar PV and wind sectors while reducing dependence on fossil fuels. However, the integration of solar and wind energy into the grid poses challenges due to their intermittent nature, especially the fluctuations in solar radiation and wind speed [7]. As a result, these energy sources alone are insufficient for ensuring a reliable power supply in microgrids.

A promising solution to address this issue is to connect renewable energy microgrids to the main grid, enabling bidirectional power flow and improving system reliability [7]. However, the costs and technical difficulties of linking remote areas to the grid remain significant challenges. A more effective strategy is the development of an Integrated Hybrid Renewable Energy System (IHRES), which combines a variety of renewable and conventional energy sources. This system also integrates multiple storage technologies, such as battery cells, fuel cells, supercapacitors, flywheels, molten salt, hydroelectric pumped storage systems, and compressed air storage, along with a backup DG to improve reliability and lower overall system costs [5,8].

This chapter focuses on optimizing the dimensions of individual components within a standalone hybrid energy system, which includes PV panels, wind turbines, DGs, and battery storage units. The analysis is based on a detailed case study of New Minya City in Egypt, utilizing current meteorological data, specifically global horizontal radiation and wind speed distribution measured at 50 m above the ground. The data are visually represented in Figures 2.1 and 2.2, providing a clear view of the environmental factors influencing the energy system. The primary objective of this research is to design an efficient and balanced configuration of the hybrid system components to ensure optimal performance and reliability [9].

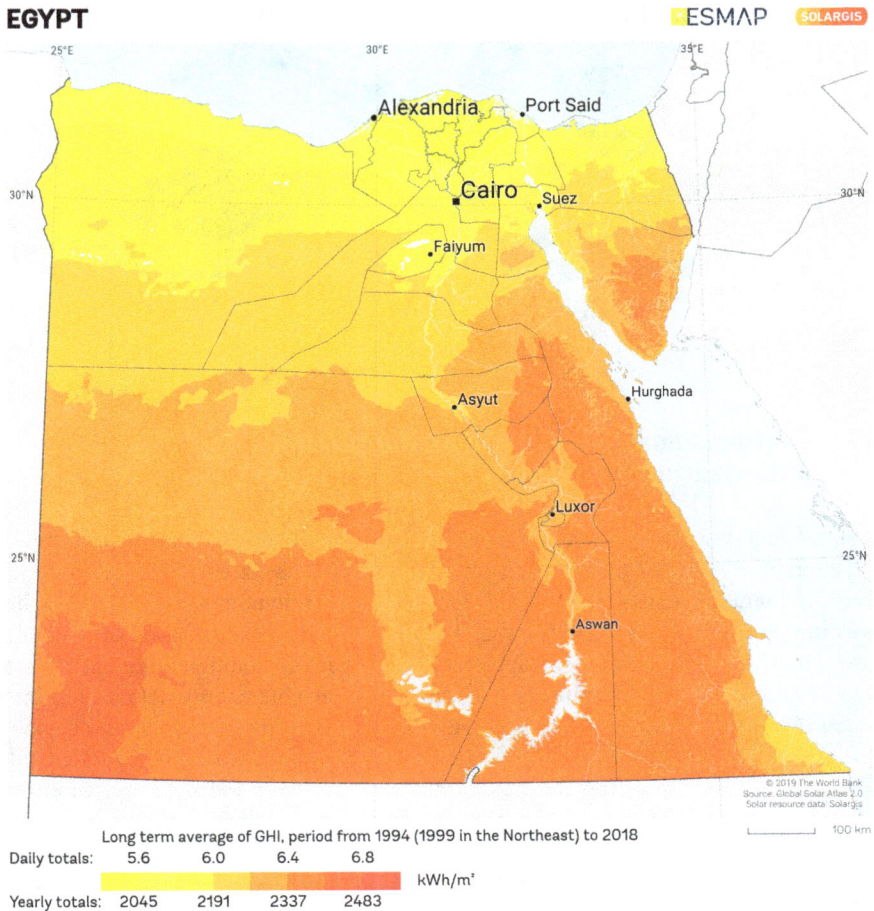

Figure 2.1   Map of global horizontal irradiation in Egypt

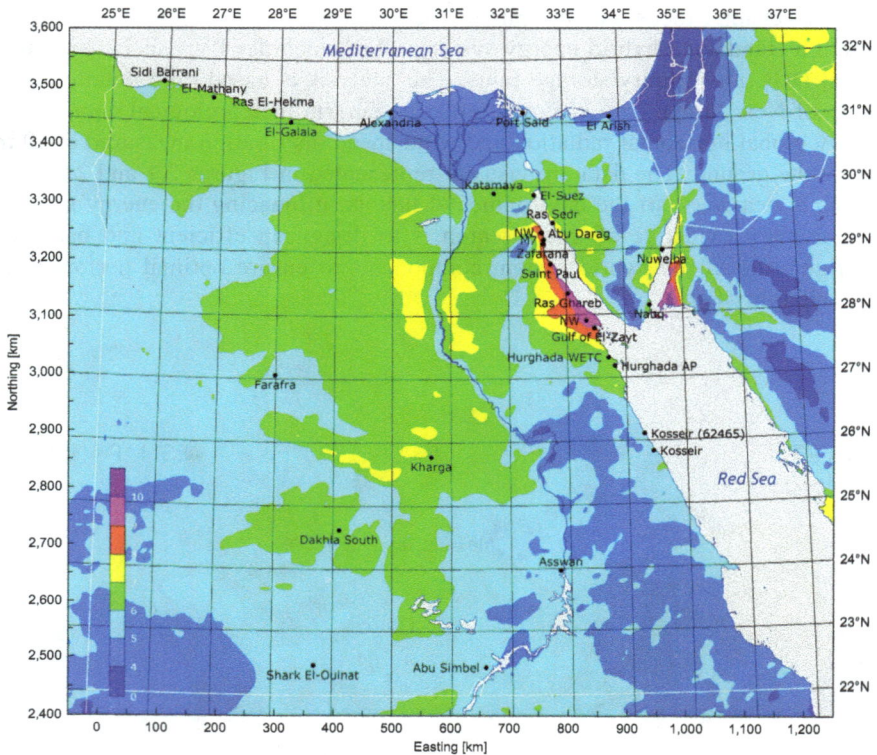

Figure 2.2   Wind speed distribution across Egypt

2.2   Comprehensive review of sizing methods and optimization techniques

A significant amount of research has focused on determining the optimal technical and economic sizing of hybrid renewable energy systems for off-grid areas. A wide array of literature exists, detailing different configurations and sizing approaches used in these hybrid systems. The majority of these systems are based on PV solar panels and wind turbines, often supplemented by backup solutions like battery storage or DGs [9–12]. In certain cases, additional renewable energy sources are integrated into the system, with excess energy stored in various energy storage devices.

Hybrid energy system sizing methodologies can be broadly categorized into three main types: (1) sizing through commercial software tools, (2) conventional sizing techniques, and (3) algorithm-driven sizing approaches.

Several commercial software tools have been developed to design and optimize hybrid energy systems, offering various features such as technical, financial, and environmental analysis. RETScreen, launched in 1998, is widely used for system sizing and optimization, considering factors like energy efficiency, system losses, and greenhouse gas emissions [13]. The software now supports the optimization of solar PV, wind, DGs, and battery storage, aiming for reductions in emissions. iHOGA, which utilizes a genetic algorithm, optimizes hybrid systems incorporating renewable and conventional energy sources, such as wind and fuel cells, and has demonstrated significant reductions in emissions and unmet load [13]. Hybrid2, developed by the Renewable Energy Research Laboratory, analyzes systems with PV, wind, and battery banks [14], while HOMER, developed by the National Renewable Energy Laboratory, is a widely used tool for both grid-connected and standalone systems, optimizing configurations to reduce costs and CO_2 emissions [14]. HybSim, focused on off-grid systems, performs techno-economic analysis for PV and distributed generation configurations [15]. TRNSYS, initially designed for thermal systems, now analyzes hybrid systems, especially those involving solar thermal and PV systems, and is useful for energy-saving strategies in buildings. Finally, Dymola supports the optimization of integrated systems, including PV, wind, and DGs, with lifecycle cost assessments for detailed system analysis [16].

Literature extensively documents traditional methods, often referred to as deterministic sizing techniques. These methods are typically categorized into four types: analytical methods, iterative methods, probabilistic methods, and artificial intelligence (AI)-based methods [9–12,15,17–20].

Analytical methods approach the hybrid system as a set of numerical equations, simplifying the system sizing problem into a mathematical function. For instance, in South Africa, this method was used to size PV panels and wind turbines, optimizing system efficiency while maintaining an energy cost of approximately €0.97 per kWh [21]. This resulted in an annual generation of around 100 GWh. Although analytical methods are computationally efficient, they tend to lack the flexibility needed to address more general optimization problems.

In contrast, iterative methods utilize an algorithm that explores various configurations through successive iterations until the optimal solution is found. This approach has been applied in Brazil to determine the sizing of PV systems, wind turbines, and batteries, resulting in a total lifecycle cost of approximately $25,672 [22]. While iterative methods are straightforward and transparent, they may fail to account for critical parameters, such as the height of wind turbine towers or the tilt angle of solar panels, which can significantly affect the overall performance of the system.

Probabilistic methods account for uncertainties, especially in wind speed and other variable inputs, reflecting the dynamic nature of renewable energy generation. While these methods are relatively easy to implement, they are often less effective in identifying the most efficient system configurations. For example, in one application, probabilistic methods were used to size a hybrid system with PV generators, wind turbines, biomass, and a battery bank, leading to a large storage capacity and higher overall costs [23].

AI techniques represent the most advanced methods for optimizing the sizing of hybrid energy systems [24]. These approaches utilize algorithms such as multi-objective self-adaptive differential evolution (MOSADE), mine blast algorithm, and preference-inspired coevolutionary algorithm [27], among others, to improve component sizing. Additional AI methods, including artificial neural networks, fuzzy logic and discrete harmony search, have also proven effective in configuring and controlling hybrid systems. These AI-based techniques employ machine learning principles to enhance the accuracy and efficiency of system sizing, offering robust solutions to complex optimization problems [25].

While commercial software tools and traditional methods are available, solving multi-objective optimization problems in hybrid systems typically requires meta-heuristic algorithms. These algorithms have gained widespread use for their ability to address complex optimization challenges with multiple objectives. One such algorithm, the strength Pareto evolutionary algorithm, has been used to optimize the sizing of PV/solar thermal/thermal distributed energy systems, achieving significant reductions in system costs and greenhouse gas emissions [26,27]. The cuckoo search algorithm has also shown improved accuracy in hybrid system sizing compared to genetic simulation and pattern search algorithms [28]. Newer metaheuristic methods, such as MOSADE, whale optimization algorithm, and moth-flame optimization, have further expanded the optimization toolset, delivering improved reductions in system costs and increased reliability [3,7,15,19,23,24,28]. These advanced methods leverage sophisticated optimization techniques to tackle the complex challenges of hybrid energy system sizing, providing more effective and efficient solutions.

2.3   Conservation energy and maintaining reliability

The primary goal of generation dispatch, or Unit Commitment, for distributed generators is to efficiently allocate generation resources, taking into account both supply availability and demand response programs. Demand-side management (DSM) aims to optimize load scheduling in a cost-effective manner, driven by customer demand response programs, with the following objectives [29]:

- Keep the load factor as close to 1.0 as possible.
- Ensure an acceptable supply margin while meeting peak demand.

Achieving these goals allows utilities to secure sufficient energy from partici-pating generation units, thereby maximizing profits and reducing the cost per kilowatt-hour [30]. To this end, traditional DSM strategies, such as peak shaving, valley filling, load shifting, and energy arbitrage, have been explored [31,32]. DSM has been applied to address energy shortages in small-scale reactive DC networks for residential and data communication applications [33]. Further solutions aimed at tackling energy deficits and stability issues arising from demand and supply fluc-tuations in small grids were proposed [34]. In rural areas, small renewable-powered grids often face challenges in responding adequately to peak loads. As a result, bat-tery energy storage systems and DGs have been recommended as effective solutions to manage peak demand. Shifting from traditional centralized DSM control to a multi-agent-based approach for more efficient balancing of supply and demand was suggested [35]. DSM implementations are generally tailored to meet specific needs. For example, peak shaving strategies have been successfully implemented, while

Energy management focuses on reliability, economics, and load scheduling through a scenario-based control scheme [30]. Saini and coworkers classified DSM into three key activities from which execution strategies can be derived [30,36]:

- Energy Demand Reduction Programs, which focus on reducing demand through improved and more efficient processes, such as smart buildings or energy-saving equipment.
- Load Management Programs, which aim to alter load patterns by shifting demand and reducing consumption during peak periods and peak rates.
- Load Growth and Conservation Programs, which focus on modifying load patterns by either replacing or delaying loads.

Energy demand management refers to programs designed to adjust customer electricity usage for both energy and cost savings. These programs typically target controlling energy consumption on the customer side of the meter, or behind the meter. Various strategies in energy demand management seek to minimize energy waste, reduce consumption during periods of low electricity supply, and lower system operating costs. The literature presents a range of strategies tailored or adapted for effective DSM. For example, a wide range of DSM strategies are explored [31], with their demand curves illustrated in Figure 2.3 [30]. These strategies aimed to

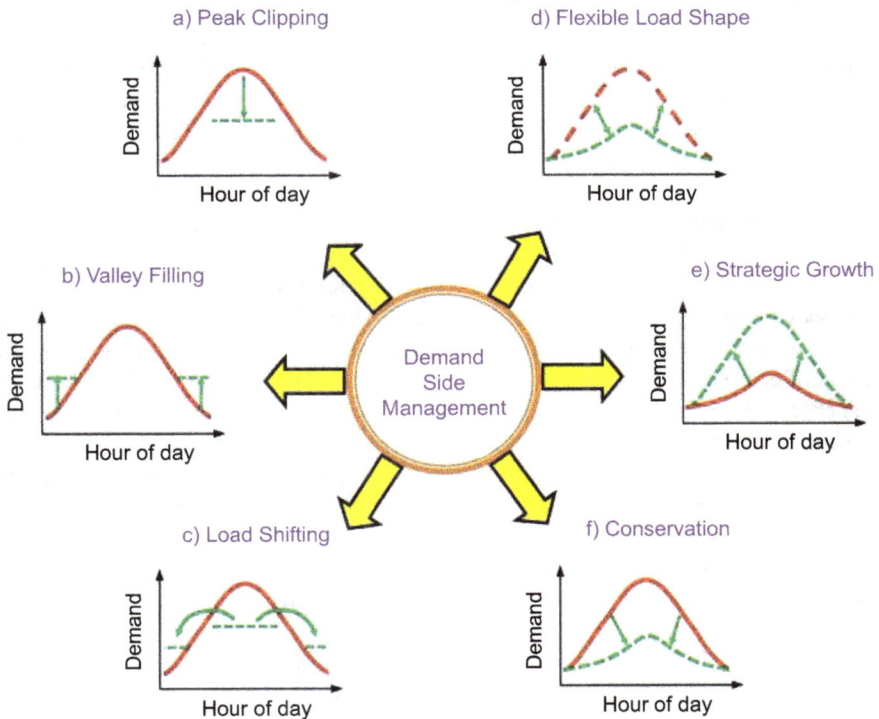

Figure 2.3   Various methods of DSM

encourage customer participation and improve the management of distributed generation, which is increasingly favored in modern energy networks [30].

2.4 DSM strategies and concepts

DSM refers to programs designed to modify customer electricity use to achieve energy and cost-related savings. DSM strategies generally focus on controlling energy consumption on the customer side of the meter, aiming to reduce energy waste, manage consumption during low supply periods, lower energy costs, and minimize system operation costs. Various strategies have been adapted or proposed in the literature for effective DSM implementation [30].

2.4.1 Peak shaving

Peak shaving, also known as peak clipping, is a time-tested DSM strategy designed to manage and reduce electricity loads during periods of peak demand, without substantially affecting the overall demand curve. This approach involves several strategic elements.

First, modification of reference loads plays a crucial role. By implementing a cap on the maximum power that can be delivered, peak power demands are significantly reduced, as demonstrated in various smart microgrid simulations. This reduction is not limited to peak power but also extends to average power consumption, which helps in managing the overall energy demand more effectively.

Second, battery power synthesis is another critical aspect of peak shaving. Optimizing the use of batteries allows for better management of energy storage and consumption. By strategically discharging batteries during peak periods and charging them during off-peak times, peak shaving can be achieved efficiently. This approach also helps in reducing the reliance on grid energy, thereby minimizing the need for additional grid buffering and enhancing the stability of the power supply.

Third, prototype control methods are essential for validating the effectiveness of peak shaving strategies. Simulations of microgrids, both in grid-connected and in island modes, are used to test and refine these methods. These simulations help in assessing the impact of peak shaving on the network and provide valuable insights into how stress on the network can be reduced. By modeling different scenarios and control strategies, it is possible to fine-tune the approach to ensure optimal performance and reliability.

Overall, peak shaving is a versatile strategy that not only helps in managing peak demand but also contributes to the overall efficiency and stability of the power grid. Through the combination of load modification, optimized battery use, and rigorous simulation testing, peak shaving can effectively address the challenges associated with peak electricity demands and support more sustainable energy management practices.

2.4.2 Valley filling

Valley filling is a strategy designed to increase energy use during periods of low demand, thereby utilizing excess generation that would otherwise go to waste. This approach optimizes the use of surplus power by shifting energy consumption to times when generation exceeds demand. Methods for implementing valley filling include employing flexible loads to concentrate energy use during off-peak times, which helps to make the most of surplus generation. Energy storage systems, such as EVs and other storage solutions, can be charged during low-demand periods and discharged to help balance demand profiles. Stochastic algorithms, such as Monte Carlo simulations, are used to stabilize the grid and effectively manage valley filling. Additionally, centralized charging strategies can be applied to smooth out demand profiles during off-peak hours, thereby improving grid stability and efficiency.

2.4.3   Load leveling

Load leveling aims to reduce fluctuations between peak and off-peak demand values, particularly in networks with large load variations. This strategy focuses on smoothing out demand profiles to enhance stability and efficiency. Key techniques include maximum power point tracking, which improves energy storage and levels out load profiles by optimizing system performance. Another method is the integration of EVs, which can be used for load leveling and providing ancillary services such as frequency regulation. Additionally, reactive power-based load leveling employs reactive power controls to stabilize load profiles and manage demand variations effectively. These strategies are essential for maintaining a steady and balanced load, crucial for the reliable operation of power distribution networks.

2.4.4   Load shifting

Load shifting involves transferring loads from peak to off-peak periods without altering overall consumption patterns. This strategy helps in balancing demand by moving energy usage to times when the grid is less stressed. Techniques employed in load shifting include day-ahead load shifting, which schedules loads for off-peak periods to alleviate peak demand [13]. Additionally, multi-objective optimization frameworks are used to reshape load profiles and reduce peak demand, allowing for more efficient energy management and grid stability [14]. These approaches help in managing demand more effectively and improving the overall efficiency of energy systems.

2.4.5   Energy arbitrage

Energy arbitrage involves storing energy during low-price periods and utilizing it when prices are higher. Efficient storage systems, such as batteries and pumped hydro, are employed to capture energy when it is inexpensive and release it during times of higher prices. For instance, vanadium redox batteries are optimized to enhance the profitability of energy arbitrage [15]. Additionally, community distribution systems use advanced techniques like adaptive neuro-fuzzy inference systems for managing voltage and performing energy arbitrage effectively [16]. These strategies demonstrate how energy arbitrage can be implemented through various storage technologies and sophisticated control systems to maximize economic benefits and improve energy management.

2.4.6   Strategic conservation

Strategic conservation is a DSM strategy focused on reducing energy consumption by encouraging more efficient use of energy or by minimizing the amount of energy services required. This consumer-centered approach involves utilities offering incentives or programs to influence customer behavior, such as rebates for energy-efficient appliances or adjustments to usage patterns through time-of-use pricing. The goal is to improve energy efficiency and reduce consumption without sacrificing comfort or service quality. Effective strategic conservation often relies on accurate demand forecasting, which predicts how electricity is consumed and informs adjustments to reduce peak demand or shift consumption patterns. Examples include load shifting, where energy use is moved to off-peak times, or load reduction during peak periods. Implementing control systems that optimize energy use based on real-time data can also be part of the strategy. Utilities benefit from reduced peak demand and operational costs, while customers can experience lower energy bills and enhanced comfort. However, balancing energy conservation with consumer comfort and ensuring that measures do not disrupt service can be challenging. Strategic conservation thus plays a vital role in achieving a more sustainable and cost-effective energy system by aligning energy usage with both economic and environmental goals.

2.4.7   Strategic load growth

Strategic load growth involves planned increases in energy sales by leveraging technologies such as EVs and automation. This strategy aims to boost utility revenue and enhance customer productivity by expanding market share and encouraging higher energy consumption. Strategic load growth often utilizes smart technologies to manage and optimize energy use, ultimately leading to increased sales and improved efficiency. By focusing on these technologies, utilities can minimize waste energy (dump energy) and reduce operational costs, contributing to overall energy system efficiency. This approach not only helps utilities grow their market share but also supports energy efficiency and cost savings through advanced management and optimization techniques.

2.4.8   Flexible load shapes

Flexible load shapes involve offering incentives for load curtailment and adjusting load profiles according to supply availability, energy costs, and customer needs. This strategy is especially beneficial in systems with intermittent renewable energy sources, as it allows for dynamic adjustment of energy consumption patterns. By aligning load profiles with the variability of renewable energy supply, flexible load shapes help to optimize energy use, reduce costs, and enhance system reliability. This approach enables utilities to manage demand more effectively and provides customers with opportunities to save on energy costs while contributing to a more stable and efficient energy system.

DSM strategies play a pivotal role in optimizing power dispatch, storage management, and overall system operations, especially in systems reliant on renewable energy and distributed generation. Optimization methods are crucial for meeting techno-economic goals such as reducing costs, improving efficiency, and minimizing energy consumption [20]. They enable better management of energy resources and facilitate the alignment of supply with demand. These strategies and concepts are essential for enhancing the efficiency of energy systems. They contribute to improved reliability, cost savings, and optimized energy use, benefiting both utilities and consumers alike.

2.5 Mathematical modeling of IHRES components

To determine the optimal configuration for an IHRES, it is essential to employ precise mathematical modeling of its components [9,11,12,19]. This research proposes a model that integrates various energy sources to address the electricity needs of New Minya City, Egypt. The model incorporates solar and wind energy, complemented by a DG for backup power and a battery storage system for energy storage and management. The proposed hybrid system is designed to economically resolve electricity shortages by combining solar PV modules, wind turbines, distributed turbines, and battery storage. The system, as depicted in Figure 2.4, comprises seven fundamental components:

1. PV Solar Modules: These units capture and convert sunlight into electricity, forming a primary source of renewable energy.
2. Wind Turbines: These devices harness wind energy, contributing an additional renewable source to the system.
3. Diesel Generators: These generators provide reliable backup power, ensuring energy availability during periods when renewable sources are insufficient.
4. Battery Storage System: This component stores excess energy produced by renewable sources, making it available for use when generation is low.
5. Bidirectional Power Converter: This device manages the flow of electricity between the various components of the system, facilitating efficient energy distribution and utilization.
6. Discharge Load: This represents the energy demand that the system meets by distributing the generated power.
7. Service Load: This refers to the basic energy requirements that the system fulfills to ensure a consistent energy supply.

By accurately modeling these components, the IHRES can be optimized to achieve a balance between energy supply and demand, enhancing the system's efficiency and reliability.

2.5.1 PV panel model

Solar panels convert solar radiation into electrical energy. The hourly electricity output of a PV panel, denoted as $P_{PV}(t)$, is determined by the amount of solar radiation

DC Bus   AC Bus

PV Panels   Wind Turbines

Battery Bank   Diesel Generator

BDC-CC

Dump Load   Load

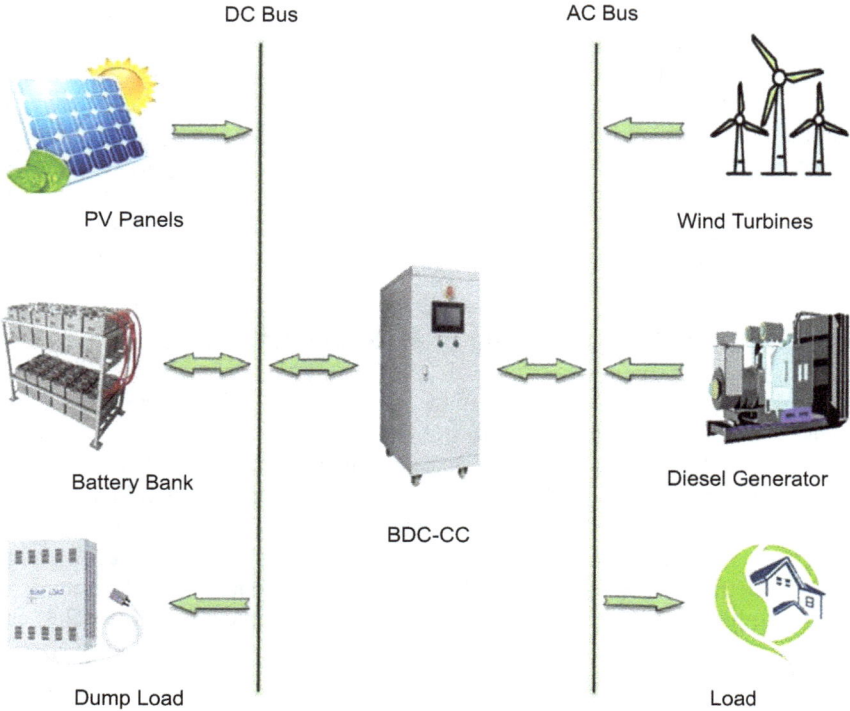

Figure 2.4   Configuration of the HRES

incident on the tilted surface of the panels, represented by $G_T(t)$. This relationship is expressed by the equation provided in reference components [9,11,12,19]

$$P_{PV}(t) = P_R \times D_F \times \left(\frac{G_T(t)}{G_{ref}}\right) \times \left[1 + \tau_p \times \left(T_C(t) - T_{ref}\right)\right] \qquad (2.1)$$

where P_R represents the rated capacity of the PV panel (in kW), while D_F is the derating factor of the PV system, set at 80% for this study. The parameter G_{ref} refers to the solar radiation under standard test conditions (STC) [4], and τ_p is the temperature coefficient of power, valued at $0.0037/°C$ in this study. T_{ref} is the reference temperature of the PV cell at STC, which is 25°C, and $T_C(t)$ denotes the cell temperature of the panel under operating conditions. The incident radiation on a tilted PV module is calculated using the equation provided by [5]

$$G_T(t) = G_b(t)R_b(t) + G_d(t)\left(\frac{1 + \cos\beta}{2}\right) + G(t)\rho_g\left(\frac{1 - \cos\beta}{2}\right) \qquad (2.2)$$

In this equation, $G_b(t)$ represents the beam radiation, $G_d(t)$ stands for the diffuse radiation, and $G(t)$ denotes the global radiation on a horizontal surface at a given hour (t). The term $R_b(t)$ is the geometry factor that accounts for the orientation of the PV panel relative to the sun, β refers to the tilt angle of the PV module, and ρ_g is the ground reflectivity. Additionally, the temperature of the PV panel's cell under operating conditions can be expressed as

$$T_A(t) = T_a(t) + \frac{G(t)}{G_{ref}} . \text{NOCT} \tag{2.3}$$

where $T_C(t)$ is the cell temperature, $T_a(t)$ is the ambient temperature, and NOCT (nominal operating cell temperature) is a factor provided by the manufacturer to estimate cell temperature at operating conditions

$$T_C(t) = T_A(t) + \left(\frac{T_{NOCT} - 20}{G_{NOCT}}\right) \times G_T(t) \tag{2.4}$$

In this equation, T_A represents the ambient temperature (°C), T_{NOCT} refers to the nominal operating cell temperature (NOCT), and G_{NOCT} is the solar radiation at NOCT, which is typically 0.80 kW/m². The total hourly electricity generation of a PV system, $E_{PV}(t)$, can be expressed as

$$E_{PV}(t) = N_{PV} \cdot P_{PV}(t) \tag{2.5}$$

In this equation, N_{PV} represents the number of PV panels in the system, and $P_{PV}(t)$ denotes the hourly electricity generation of a single PV panel at time (t). This formula accounts for the combined output of all PV panels in the system, where the generation from each panel is influenced by factors such as solar irradiance, temperature, and system efficiency. MATLAB Code 2.1 presents the implementation of the calculation of the PV power generation.

```
% Define parameters
P_R = 0.4; % Rated capacity of the PV panel in kW (400 W)
D_F = 0.80; % Derating factor
G_ref = 1.0; % Solar radiation under standard test
conditions (STC) in kW/m2
tau_p = 0.0037; % Temperature coefficient of power per °C
T_ref = 25; % Reference temperature of the PV cell at STC
in °C
NOCT = 45; % Nominal Operating Cell Temperature in °C
G_NOCT = 0.80; % Solar radiation at NOCT in kW/m2
beta = 30; % Tilt angle of the PV module in degrees
rho_g = 0.2; % Ground reflectivity
N_PV = 10; % Number of PV panels
```

MATLAB Code 2.1   Calculation of the PV power generation.

```matlab
% Time vector (example for one day in hours)
t = 0:23; % Hours of the day

% Example input data for solar radiation and ambient
temperature
G_b = rand(1,24) * 0.6; % Beam radiation in kW/m2
G_d = rand(1,24) * 0.3; % Diffuse radiation in kW/m2
G = G_b + G_d; % Global radiation on a horizontal surface
in kW/m2
T_a = 20 + 5 * sin(pi * t / 12); % Ambient temperature in
°C (example sinusoidal variation)

% Calculate the incident radiation on the tilted PV
module
G_T = G_b .* cosd(beta) + G_d .* ((1 + cosd(beta)) / 2) +
G .* rho_g .* ((1 - cosd(beta)) / 2);

% Calculate the cell temperature T_C
T_A = T_a + (G / G_NOCT) .* NOCT;
T_C = T_A + ((NOCT - 20) / G_NOCT) .* G_T;

% Calculate the hourly electricity output of a single PV
panel
P_PV = P_R * D_F .* (G_T / G_ref) .* (1 + tau_p .* (T_C -
T_ref));

% Ensure that the power output is non-negative
P_PV(P_PV < 0) = 0;

% Total hourly electricity generation of the PV system
E_PV = N_PV * P_PV; % kW

% Display results
figure;
subplot(3,1,1);
plot(t, G_T, 'r', 'LineWidth', 1.5);
xlabel('Hour of the Day');
ylabel('Incident Radiation (G_T) [kW/m^2]');
title('Incident Radiation on Tilted PV Module');

subplot(3,1,2);
plot(t, T_C, 'b', 'LineWidth', 1.5);
xlabel('Hour of the Day');
ylabel('Cell Temperature (T_C) [°C]');
title('Cell Temperature of PV Panel');

subplot(3,1,3);
plot(t, E_PV, 'g', 'LineWidth', 1.5);
xlabel('Hour of the Day');
```

MATLAB Code 2.1   (Continued)

```
ylabel('Electricity Output (E_PV) [kW]');
title('Hourly Electricity Output of PV System');

% Output the total daily energy generation
total_energy = sum(E_PV); % kW
disp(['Total Daily Energy Generation: ',
num2str(total_energy), ' kWh']);
```

MATLAB Code 2.1 (Continued)

2.5.2 Wind energy system model

The wind energy system model is influenced by factors such as wind speed at hub height, wind turbine (WT) speed characteristics, and the surface type of the location components [9,11,12,19]. The wind speed at hub height $(u(h))$ can be calculated based on the wind speed measured at anemometer height $(u(h_a))$ using the following equation:

$$u(h) = u(h_a)\left(\frac{h}{h_a}\right) \tag{2.6}$$

where α is the roughness factor, which varies depending on the location and is typically 0.20 under normal wind conditions and 0.11 in strong winds.

The mechanical output power of a WT (P_{mech}) is given by

$$P_{mech} = \frac{1}{2}\rho A C_p u^3 \tag{2.7}$$

where A is the rotor blade swept area, u is the wind speed, C_p is the power coefficient (with a theoretical maximum of 0.593), and ρ is the air density. The turbine's tip-speed ratio λ is calculated as

$$\lambda = \frac{\omega R}{u} \tag{2.8}$$

where R is the turbine's radius, and ω is the angular velocity.

Wind turbines begin generating power at the "cut-in" wind speed and stop at the "cut-off" speed to avoid rotor damage. The rated power output is achieved between these speeds. The power delivered by the turbine (P_{WT}) is described by a piecewise function that accounts for different wind speed ranges

$$P_{WT} = \begin{cases} 0 & u(t) < u_{cut-in} \\ \dfrac{N_{WT} * P_{R_{WT}} * (u^2(t) - u^2_{cut-in})}{u^2_{rated} - u^2_{cut-in}} & u_{cut-in} \le u(t) \le u_{rated} \\ N_{WT} * P_{RWT} & u_{rated} < u(t) \le u_{cut-off} \\ 0 & u(t) > u_{cut-off} \end{cases} \tag{2.9}$$

where N$_{WT}$ is the number of turbines, $u_{cut\text{-}in}$, u_{rated}, and $u_{cut\text{-}off}$ are the respective wind speeds, and $P_{R_{WT}}$ is the rated output power of the WT. This model accounts for the different phases of WT operation, ensuring safe and efficient energy generation. The implementation using MATLAB has been MATLAB Code 2.2.

```matlab
% Define parameters
h_a = 10; % Anemometer height in meters
h = 50; % Hub height of the wind turbine in meters
alpha = 0.20; % Roughness factor
rho = 1.225; % Air density in kg/m^3 (sea level at 15°C)
A = 500; % Rotor blade swept area in m^2 (example value)
C_p = 0.45; % Power coefficient (assumed typical value)
R = 30; % Radius of the turbine blades in meters
omega = 10; % Angular velocity in rad/s (example value)
u_cut_in = 3; % Cut-in wind speed in m/s
u_rated = 12; % Rated wind speed in m/s
u_cut_off = 25; % Cut-off wind speed in m/s
P_R_WT = 2; % Rated power output of the wind turbine in
MW
N_WT = 5; % Number of wind turbines

% Time vector (example for one day in hours)
t = 0:23; % Hours of the day

% Example wind speed data measured at anemometer height
(u(h_a))
u_h_a = 5 + 5 * sin(pi * t / 12); % Example wind speed
data in m/s

% Calculate wind speed at hub height (u(h))
u_h = u_h_a .* (h / h_a) .^ alpha;

% Calculate mechanical output power of the wind turbine

P_mech = 0.5 * rho * A * C_p * u_h .^ 3; % Mechanical
power in W

% Calculate tip-speed ratio (lambda)
lambda = (omega * R) ./ u_h;

% Calculate power delivered by the wind turbine (P_WT)
based on wind speed ranges
P_WT = zeros(size(u_h)); % Initialize power output vector
for i = 1:length(u_h)
    if u_h(i) < u_cut_in
        P_WT(i) = 0;
    elseif u_h(i) <= u_rated
        P_WT(i) = (P_R_WT / (u_rated - u_cut_in)) *
(u_h(i) - u_cut_in)^2;
```

MATLAB Code 2.2   Implementation of wind energy calculations

```
      elseif u_h(i) <= u_cut_off
          P_WT(i) = P_R_WT;
      else
          P_WT(i) = 0;
      end
end

% Total hourly electricity generation of the wind system
E_WT = N_WT * P_WT / 1000; % Convert to MW

% Display results
figure;
subplot(3,1,1);
plot(t, u_h, 'r', 'LineWidth', 1.5);
xlabel('Hour of the Day');
ylabel('Wind Speed at Hub Height (u(h)) [m/s]');
title('Wind Speed at Hub Height');

subplot(3,1,2);
plot(t, P_mech / 1e6, 'b', 'LineWidth', 1.5); % Convert
to MW
xlabel('Hour of the Day');
ylabel('Mechanical Output Power (P_m_e_c_h) [MW]');
title('Mechanical Output Power of Wind Turbine');

subplot(3,1,3);
plot(t, E_WT, 'g', 'LineWidth', 1.5);
xlabel('Hour of the Day');
ylabel('Electricity Output (E_WT) [MW]');
title('Hourly Electricity Output of Wind System');

% Output the total daily energy generation
total_energy_WT = sum(E_WT); % MW
disp(['Total Daily Energy Generation: ',
num2str(total_energy_WT), ' MWh']);
```

MATLAB Code 2.2 (Continued)

2.5.3 *Battery storage model*

Integrating a battery storage unit is crucial to improving the reliability of standalone renewable energy systems. This storage component captures surplus electricity produced by renewable sources, such as solar panels or wind turbines, making it available for times when generation falls short of meeting the load demand. During operation, electricity is prioritized to fulfill the immediate load, while any excess energy is directed to the battery banks for storage. The charging and discharging cycles of the battery system are managed by specific models, as outlined in components [9,11,12,19], which optimize the balance between energy supply and demand under varying operational conditions

$$B_{SOC}(t) = B_{SOC}(t-1) \times (1-\sigma) + \left[E_{PV}(t) + \frac{E_{WT}(t)}{\eta_C} - \frac{E_{Load}(t)}{\eta_C} \right] \times \eta_B$$

(2.10)

$$B_{SOC}(t) = B_{SOC}(t-1) \times (1-\sigma) - \left[\frac{E_{Load}(t)}{\eta_C} - E_{PV}(t) - \frac{E_{WT}(t)}{\eta_C} \right]$$ (2.11)

The battery capacity at any given hour, $B_{SOC}(t)$, is influenced by several factors, including the battery's state of charge from the previous hour, $B_{SOC}(t-1)$, the self-discharging rate (σ) (set at 0.16% per year), and the charging efficiency (η_B) of the battery [7,28,37]. These parameters are critical for accurately modeling the battery's charge and discharge cycles, ensuring efficient energy storage and availability. The MATLAB code implementation may be written as in MATLAB Code 2.3.

```
% Define parameters
initial_SOC = 0.5; % Initial state of charge (e.g., 50%)
battery_capacity_Ah = 300; % Battery capacity in Ah
voltage = 12; % Battery voltage in V
self_discharging_rate = 0.16 / 100 / 365; % Self-
discharging rate per hour
charging_efficiency = 0.9; % Charging efficiency
discharging_efficiency = 0.9; % Discharging efficiency
DC_AC_conversion_efficiency = 0.95; % DC to AC conversion
efficiency
AC_DC_conversion_efficiency = 0.95; % AC to DC conversion
efficiency
cost_per_kWh = 700; % Initial capital cost per kWh in $
OM_cost_per_kWh = 10; % Annual O&M cost per kWh in $

% Time vector (example for one day in hours)
t = 0:23; % Hours of the day

% Example data for solar, wind, and load (in kWh)
E_PV_DC = 5 * sin(pi * t / 12) + 5; % Example solar
energy generation in kWh (DC)
E_WT_AC = 4 * sin(pi * t / 12); % Example wind energy
generation in kWh (AC)
E_Load_AC = 4 + 2 * sin(pi * t / 24); % Example load
demand in kWh (AC)
```

MATLAB Code 2.3   Battery MATLAB code

```matlab
% Initialize state of charge (SOC) vector
B_SOC = zeros(size(t));
B_SOC(1) = initial_SOC * battery_capacity_Ah * voltage /
1000; % Initial SOC in kWh

% Initialize energy balance variables
E_PV_AC = zeros(size(t)); % Energy from PV converted to
AC
E_WT_DC = zeros(size(t)); % Energy from Wind converted to
DC
E_Load_DC = zeros(size(t)); % Load demand in DC
charge_energy = zeros(size(t)); % Energy added to the
battery
discharge_energy = zeros(size(t)); % Energy taken from
the battery

% Calculate SOC and energy balance for each hour
for i = 2:length(t)
    % Convert PV energy from DC to AC
    E_PV_AC(i) = E_PV_DC(i) *
DC_AC_conversion_efficiency;

    % Convert wind energy from AC to DC
    E_WT_DC(i) = E_WT_AC(i) *
AC_DC_conversion_efficiency;

    % Convert load demand from AC to DC
    E_Load_DC(i) = E_Load_AC(i) /
AC_DC_conversion_efficiency;

    % Total DC power available from PV and wind
    Total_DC_Power = E_PV_DC(i) + E_WT_DC(i);

    % Battery charging/discharging
    if Total_DC_Power > E_Load_DC(i)
        % Battery is charging
        charge_energy(i) = (Total_DC_Power -
E_Load_DC(i)) * charging_efficiency; % Energy added to
battery
        B_SOC(i) = B_SOC(i-1) * (1 -
self_discharging_rate) + charge_energy(i);
    else
        % Battery is discharging
        discharge_energy(i) = (E_Load_DC(i) -
Total_DC_Power) / discharging_efficiency; % Energy taken
```

MATLAB Code 2.3   (Continued)

```
from battery
        B_SOC(i) = B_SOC(i-1) * (1 -
self_discharging_rate) - discharge_energy(i);
    end

    % Ensure SOC does not exceed battery capacity limits
    B_SOC(i) = min(max(B_SOC(i), 0), battery_capacity_Ah
* voltage / 1000);
end

% Convert SOC from kWh back to Ah for reporting
B_SOC_Ah = B_SOC * 1000 / voltage;

% Calculate costs
initial_cost = battery_capacity_Ah * voltage *
cost_per_kWh / 1000; % Initial capital cost in $
OM_cost = battery_capacity_Ah * voltage * OM_cost_per_kWh
/ 1000; % Annual O&M cost in $

% Plot results
figure;
subplot(3,1,1);
plot(t, B_SOC, 'b', 'LineWidth', 1.5); % SOC in kWh
xlabel('Hour of the Day');
ylabel('State of Charge (B_SOC) [kWh]');
title('Battery State of Charge');

subplot(3,1,2);
plot(t, B_SOC_Ah, 'g', 'LineWidth', 1.5); % SOC in Ah
xlabel('Hour of the Day');
ylabel('State of Charge (B_SOC) [Ah]');
title('Battery State of Charge in Ampere-Hours');

subplot(3,1,3);
plot(t, E_PV_AC, 'r', 'LineWidth', 1.5); hold on;
plot(t, E_WT_AC, 'm', 'LineWidth', 1.5); hold on;
plot(t, E_Load_AC, 'k', 'LineWidth', 1.5);
xlabel('Hour of the Day');
ylabel('Energy [kWh]');
title('Energy Balance');
legend('PV Energy (AC)', 'Wind Energy (AC)', 'Load Demand
(AC)');

% Output costs
disp(['Initial Capital Cost: $', num2str(initial_cost)]);
disp(['Annual O&M Cost: $', num2str(OM_cost)]);
```

MATLAB Code 2.3   (Continued)

2.5.4 DG backup system model

In isolated grid systems, internal combustion DGs are invaluable as they offer a reliable source of backup power during periods when renewable energy sources, such as solar or wind, are unavailable. This can occur due to extended periods of cloud cover, prolonged rainy seasons, or situations where battery storage systems are depleted and unable to meet the electricity demand [7,9–11]. DGs ensure uninterrupted power supply by stepping in to bridge the gap, maintaining system reliability and operational stability. The fuel consumption of a DG, denoted as F_{DG}, is primarily influenced by the load demand it needs to support. This demand varies throughout the day, depending on energy consumption patterns. The relationship between the required load and the corresponding fuel consumption can be expressed through a linear model. This model provides an effective means to estimate fuel use based on the generator's output, optimizing the DG's operational efficiency and ensuring fuel is used as economically as possible. By calculating fuel consumption in this manner, system operators can better manage resources, control operational costs, and ensure sustainable energy generation even during periods of renewable energy shortfall. The hourly fuel consumption of a DG can be written as

$$F_{DG}(t) = \left[a_{DG} * P_{DGGen}(t) + b_{DG} * P_{DGrating} \right] (l/h) \tag{2.12}$$

The fuel consumption of a DG is characterized by its fuel consumption curve, which depends on the generator's rated power and its actual hourly generated power. The coefficients a_{DG} and b_{DG} used in this model are crucial for accurately estimating fuel usage. For the DG, these coefficients are $a_{DG} = 0.246$ l/kWh and $b_{DG} = 0.08145$ l/kWh.

In this context, $P_{DG_{rating}}$ represents the DG's rated power capacity, while $P_{DG_{Gen}(t)}$ denotes the power generated by the DG during a specific hour. The fuel consumption calculation utilizes these coefficients to provide a reliable estimate of the fuel needed based on the power output, ensuring efficient operation and cost management for the generator. The DG (AFC)'s annual fuel consumption is calculated using the formulation of

$$AFC = \sum_{t=1}^{8760} F_{DG}(t) \tag{2.13}$$

2.5.5 CO₂ emissions

Estimates suggest that the hourly fuel consumption and corresponding CO_2 emissions of a DG can be calculated using the following formula [9,19,38]:

$$CO_2(t) = SE_{CO_2} \times F_{DG}(t) \tag{2.14}$$

In this equation, SE_{CO_2} represents the specific CO_2 emissions per liter of diesel, with a value of 2.7 kg/l. $F_{DG}(t)$ denotes the hourly fuel consumption of DG in liters per hour.

To estimate the DG's annual CO_2 emissions, the total yearly fuel consumption must be multiplied by the specific CO_2 emission factor. This approach provides a comprehensive view of the generator's environmental impact over a year, helping to assess its contribution to greenhouse gas emissions and guide decisions on fuel management and emission reduction strategies.

$$\text{Annual}_{CO_2}\text{\_Emissions} = \sum_{t=1}^{8760} CO_2(t) \qquad (2.15)$$

The MATLAB code for the calculation of the DG and its fuel consumption can be shown in MATLAB Code 2.4.

```
% Define Parameters
a_DG = 0.246; % Coefficient for DG fuel consumption
(1/kWh)
b_DG = 0.08145; % Coefficient for DG fuel consumption
(1/kWh)
P_DG_rating = 500; % DG rated power capacity (kW)
SE_CO2 = 2.7; % Specific CO2 emissions per liter of
diesel (kg/l)

% Define sample time steps (hours in a year, 8760 hours)
time_steps = 1:8760;
% Define power generated by DG (P_DG_Gen) for each hour
% For demonstration purposes, let's use random values
% Replace this with actual DG power data
P_DG_Gen = 0.5 * P_DG_rating + 0.5 * rand(1, 8760) *
P_DG_rating;
% Initialize arrays
F_DG = zeros(size(time_steps)); % Fuel consumption (1/h)
CO2 = zeros(size(time_steps)); % CO2 emissions (kg)
% Calculate hourly fuel consumption and CO2 emissions
for t = time_steps
    F_DG(t) = a_DG * P_DG_Gen(t) + b_DG * P_DG_rating;
    CO2(t) = SE_CO2 * F_DG(t);
end
% Calculate annual fuel consumption
AFC = sum(F_DG); % Annual Fuel Consumption (liters)
% Calculate annual CO2 emissions
Annual_CO2_Emissions = sum(CO2); % Annual CO2 Emissions
(kg)
% Display results
fprintf('Annual Fuel Consumption: %.2f liters\n', AFC);
fprintf('Annual CO2 Emissions: %.2f kg\n',
Annual_CO2_Emissions);
% Plot results for visualization
figure;
subplot(2, 1, 1);
plot(time_steps, F_DG, 'b');
title('Hourly Fuel Consumption of DG');
xlabel('Time (hours)');
ylabel('Fuel Consumption (1/h)');
subplot(2, 1, 2);
plot(time_steps, CO2, 'r');
title('Hourly CO2 Emissions of DG');
xlabel('Time (hours)');
ylabel('CO2 Emissions (kg)');
```

MATLAB Code 2.4   Calculation of the DG power and its fuel consumption

2.5.6 *Bidirectional or dual converter with charge controller (BDC-CC) model*

The BDC-CC is a critical component in managing energy flows in renewable energy systems. It operates in two primary modes: rectifier mode and inverter mode. In rectifier mode, the BDC-CC converts alternating current (AC) to direct current (DC), while in inverter mode, it converts DC to AC. To ensure the battery bank remains within safe operating limits, a charge controller is integrated to prevent overcharging and over-discharging [9,19,38].

The power rating of the BDC-CC (P_{BDC-CC}) varies depending on its operating mode and the conditions of power supply and demand. The following formulas are used to calculate the power rating of the BDC-CC:

1. When WT power exceeds load demand:
    If the power generated by the WT ($P_{WT}(t)$) is greater than the load demand ($P_{Load}(t)$), or if the combined power from both wind turbines and PV panels significantly exceeds the load demand ($P_{WT}(t) + P_{PV}(t) \gg P_{Load}(t)$), and the battery is not yet fully charged ($E_{Bat}(t) < E_{Bat_{max}}$), the BDC-CC starts charging the battery. The power rating of the BDC-CC in this scenario is calculated as

    $$P_{BDC-CC}(t) = [P_{WT}(t) - P_{Load}(t)] \times \eta_{Conv} \qquad (2.16)$$

2. When WT power is less than load demand:
    If the WT produces less power than the load demand ($P_{WT}(t) < P_{Load}(t)$), but the combined power from wind turbines and PV panels exceeds the load demand ($P_{WT}(t) + P_{PV}(t) > P_{Load}(t)$) and the battery is still charging ($E_{Bat}(t) < E_{Bat_{max}}$), a dump load is created if the battery is completely charged and the surplus power is directed to it. The BDC-CC power rating in this case is

    $$P_{BDC-CC}(t) = [P_{PV}(t) - P_{Load}(t)] \times \eta_{Conv} \qquad (2.17)$$

3. When both WT and PV panel power are insufficient:
    If the combined power from wind turbines and PV panels ($P_{WT}(t) + P_{PV}(t)$) is less than the load demand ($P_{Load}(t)$), and the battery is not fully discharged ($E_{Bat}(t) > E_{Bat_{min}}$), the BDC-CC provides power from the battery to meet the deficit. The power rating for the BDC-CC in this scenario is

    $$P_{BDC-CC}(t) = [P_{PV}(t) + P_{BD}(t)] \times \eta_{Conv} \qquad (2.18)$$

The calculation of the necessary power rating for the BDC-CC can be done by using the presented equations under diverse operational conditions, which will ensure optimal performance and efficient energy management in renewable energy systems. The MATLAB code for the implementation of the provided mathematical model can be found in MATLAB Code 2.5.

```
% Define Parameters
eta_Conv = 0.9; % Efficiency of the converter
E_Bat_max = 1000; % Maximum battery energy storage (kWh)
E_Bat_min = 200; % Minimum battery energy storage (kWh)
DOD = 0.2; % Depth of Discharge
SOC_Bat_max = 1 - DOD; % Maximum State of Charge
SOC_Bat_min = SOC_Bat_max - DOD; % Minimum State of
Charge

% Define sample time steps
time_steps = 1:24; % For simplicity, assuming 24-hour
data
P_WT = [500 600 550 400 350 300 400 450 500 600 700 800
750 650 600 550 500 450 400 350 300 250 200 150]; % Wind
turbine power (kW)
P_PV = [100 120 140 160 180 200 220 240 260 280 300 320
340 360 380 400 420 440 460 480 500 520 540 560]; % PV
power (kW)
P_Load = [400 450 500 550 600 650 700 750 800 850 900 950
900 850 800 750 700 650 600 550 500 450 400 350]; % Load
demand (kW)
E_Bat = 600; % Initial battery energy (kWh)

% Initialize BDC-CC power ratings
P_BDC_CC = zeros(size(time_steps));

% Loop through each time step
for t = time_steps
    % Case 1: Wind turbine power exceeds load demand
    if P_WT(t) > P_Load(t) && (P_WT(t) + P_PV(t)) >
P_Load(t) && E_Bat < E_Bat_max
        P_BDC_CC(t) = (P_WT(t) - P_Load(t)) * eta_Conv;
    % Case 2: Wind turbine power is less than load demand
    elseif P_WT(t) < P_Load(t) && (P_WT(t) + P_PV(t)) >
P_Load(t) && E_Bat < E_Bat_max
        P_BDC_CC(t) = (P_PV(t) - P_Load(t)) * eta_Conv;
    % Case 3: Both wind turbine and PV panel power are
insufficient
    elseif (P_WT(t) + P_PV(t)) < P_Load(t) && E_Bat >
E_Bat_min
        P_BDC_CC(t) = (P_PV(t) + E_Bat - P_Load(t)) *
eta_Conv;
    end
end
% Display the results
disp('BDC-CC Power Ratings (kW):');
disp(P_BDC_CC);
```

MATLAB Code 2.5   MATLAB code for implementation of the mathematical model

2.5.7 *System reliability*

Reliability refers to the capability of a power system to provide energy consistently over a specified period and under defined conditions. In this study, the reliability of the IHRES is evaluated using the LPSP. The LPSP is determined by dividing the total number of hours with power outages by the total hourly energy demands. To calculate the LPS at any given hour (t), the following formula is used [9,19,38]:

$$LPS(t) = \frac{P_{Load}(t)}{\eta_{Conv}} - P_{Gen}(t) - [(1 - \sigma) * E_{Bat}(t+1) - E_{Bat}(t)] * \eta_{BD}$$

(2.19)

The LPSP has been determined using the equation provided in [134]

$$LPSP = \frac{\sum_{t=1}^{8760} LPS(t)}{\sum_{t=1}^{8760} P_{Load}(t)}$$

(2.20)

Moreover, the LPSP can be calculated using MATLAB Code 2.6.

```
% Define Parameters
eta_Conv = 0.9; % Efficiency of the converter
eta_BD = 0.95; % Efficiency of the battery discharging
sigma = 0.1; % Battery charge retention factor (example
value)

% Define sample time steps (hours in a year, 8760 hours)
time_steps = 1:8760;

% Define load demand (P_Load) and power generated (P_Gen)
for each hour
% For demonstration purposes, let's use random values
% Replace these with actual data
P_Load = 0.5 * rand(1, 8760) * 100 + 100; % Example load
demand in kW
P_Gen = 0.5 * rand(1, 8760) * 100 + 100; % Example power
generated in kW

% Define battery energy storage at each hour
% For demonstration purposes, let's use random values
% Replace these with actual data
E_Bat = 0.5 * rand(1, 8760) * 100 + 50; % Example battery
energy in kWh

% Initialize arrays
LPS = zeros(size(time_steps)); % Loss of Power Supply
LPSP = 0; % Loss of Power Supply Probability
```

MATLAB Code 2.6   LPSP MATLAB code

```matlab
% Calculate Loss of Power Supply (LPS) for each hour
for t = time_steps
    if t < 8760 % Ensure the index is within bounds
        E_Bat_next = E_Bat(t + 1); % Energy storage at
the next hour
    else
        E_Bat_next = E_Bat(t); % Use current hour value
for the last hour
    end

    % Calculate Loss of Power Supply
    LPS(t) = (P_Load(t) / eta_Conv) - P_Gen(t) - ((1 -
sigma) * E_Bat_next - E_Bat(t)) * eta_BD;
end

% Calculate Loss of Power Supply Probability (LPSP)
LPS_sum = sum(LPS(LPS > 0)); % Sum of positive LPS values
(power supply loss)
P_Load_sum = sum(P_Load); % Total load demand

LPSP = LPS_sum / P_Load_sum; % Loss of Power Supply
Probability

% Display results
fprintf('Loss of Power Supply Probability (LPSP):
%.4f\n', LPSP);

% Plot results for visualization
figure;
plot(time_steps, LPS, 'r');
title('Hourly Loss of Power Supply (LPS)');
xlabel('Time (hours)');
ylabel('Loss of Power Supply (kW)');
```

MATLAB Code 2.6   *(Continued)*

2.6   Economic analysis of the off-grid IHRES

The economic sustainability of the IHRES is assessed through several methods, including net present cost, annual levelized cost, life cycle cost (LCC), and payback period. Among these, the LCC technique is often preferred for its comprehensive representation of project costs throughout its entire lifecycle. This study calculates the LCC of the IHRES by summing all relevant costs, such as those for system component replacement, installation, initial capital, operation and maintenance (O&M), and fuel [9,19,38].

2.6.1 Cost of energy

One key metric in the economic evaluation of IHRES is the cost of energy (COE), which is widely used. The COE can be computed using the following formula [18]:

$$COE = \frac{LCC * CRF(i, T)}{\sum_{t=1}^{8760} P_{Load}(t)} \qquad (2.21)$$

where T represents the project lifetime in years, which is set to 25 years in this study. i is the real net interest rate, and CRF is the capital recovery factor, which is calculated using the following formula [8]:

$$CRF(i, T) = \frac{i(1 + i)^T}{(1 + i)^T - 1} \qquad (2.22)$$

where i is the real net interest rate and T is the project lifetime in years.

2.6.2 Life cycle cost

The LCC of the overall project can be estimated using the following formula:

$$LCC = C_{Initial\_Capital} + C_{O\&M} + C_{Rep} + C_{Fuel} - V_{Scarp} \qquad (2.23)$$

where

- $C_{Initial\ Capital}$ represents the initial capital costs,
- $C_{O\&M}$ denotes the operation and maintenance costs,
- C_{Rep} accounts for replacement costs,
- C_{Fuel} covers the fuel costs,
- V_{Scrap} is the salvage value of the components of the IHRES.

2.6.3 Initial capital costs

The initial capital costs $C_{Initial\ Capital}$ encompass expenses related to installation, civil works, electrical testing, and commissioning. In this study, it is assumed that installation and civil works account for 20% of the WT system costs and 40% of the solar system costs, respectively [16]. The initial capital cost can be estimated using the following formula:

$$C_{Initial\ Capital} = C_{WT} \times P_{R_{WT}} \times N_{WT} + C_{PV} \times PV_{panel\ rating} \times N_{PV} + C_{Bat}$$

$$\times S_{Bat_{rating}} \times N_{Bat} + C_{DG} \times P_{DG_{rating}} \times N_{DG} + C_{BDC-CC}$$

$$\times P_{BDC-CC}$$

$$(2.24)$$

where

- C_{WT} is the cost per kilowatt of the WT, including civil works (\$/kW),
- $P_{R_{WT}}$ represents the rated power output of the wind turbines,
- N_{WT} is the number of wind turbines,
- C_{PV} is the cost per kilowatt of PV panels, including civil works (\$/kW),
- $PV_{\text{panel rating}}$ denotes the rated power of the PV panels,
- N_{PV} is the number of PV panels,
- C_{Bat} is the cost per kilowatt of the energy storage batteries (\$/kW),
- $S_{Bat_{\text{rating}}}$ is the rating of the battery bank,
- N_{Bat} is the number of battery cells,
- C_{DG} is the cost of the DG system,
- $P_{DG_{\text{rating}}}$ is the rated power output of the diesel generators,
- N_{DG} is the number of diesel generators,
- C_{BDC-CC} is the cost of the bidirectional converter with charge controller (\$/kW),
- P_{BDC-CC} is the power rating of the bidirectional converter with a charge controller.

2.6.4   *Operation and maintenance costs*

The O&M costs $C_{O\&M}$ have been estimated based on an in-depth review of various research studies [9,19,38]. These costs can be calculated using the following formula:

$$C_{O\&M} = \sum_{j=1}^{T} C_{O\&M}(1) \times \left(\frac{1}{(1+i)^j} \right) \tag{2.25}$$

In this formula, $C_{O\&M}(1)$ represents the operating and maintenance expenses in the first year, and i is the discount rate.

Alternatively, $C_{O\&M}$ can be calculated using

$$C_{O\&M} = C_{O\&M_{WT}} \times T_{WT} + C_{O\&M_{PV}} \times T_{PV} + C_{O\&M_{Bat}} \times T_{Bat} + C_{O\&M_{DG}}$$
$$\times T_{DG} + C_{O\&M_{BDC-CC}} \times T_{BDC-CC}$$

$$\tag{2.26}$$

Here

- $C_{O\&M_{WT}}, C_{O\&M_{PV}}, C_{O\&M_{Bat}}, C_{O\&M_{DG}},$ and $C_{O\&M_{BDC-CC}}$ are the costs associated with the operation and maintenance of wind turbines, PV panels, battery storage systems, diesel generators, and bidirectional converters, respectively.
- $T_{WT}, T_{PV}, T_{Bat}, T_{DG},$ and T_{BDC-CC} represent the operating time for each of these components.

2.6.5   Replacement cost

To calculate the present value of the replacement costs (C_{Rep}) for the hybrid system components over the project's lifespan, use the following formula:

$$C_{Rep} = \sum_{j=1}^{N_{Rep}} \left[K_{Rep} \times C_u \times \left(\frac{1}{1+i} \right)^{T \times \frac{j}{N_{Rep}+1}} \right] \tag{2.27}$$

where

- K_{Rep} represents the capacity of the replacement components (in kW for wind turbines, PV panels, diesel generators, and bidirectional converters, and in kWh for batteries).
- C_u is the cost of the replacement components (in \$/kW for wind turbines, PV panels, diesel generators, and bidirectional converters, and in \$/kWh for batteries).
- N_{Rep} is the number of replacements anticipated over the project's lifespan (T).
- i is the discount rate.

2.6.6   Fuel cost

The fuel cost (C_{Fuel}) can be calculated using the following formula:

$$C_{Fuel} = \left(\sum_{t=1}^{8760} F_{DG}(t) \right) \times P_{Fuel} \tag{2.28}$$

where

- $\sum_{t=1}^{8760} F_{DG}(t)$ represents the total annual fuel consumption of the DG (in liters).
- P_{Fuel} is the price of fuel per liter. In this study, the price is assumed to be \$0.80 per liter.

2.6.7   Scrap value

To determine the scrap value (V_{Scrap}), the following method can be used:

$$V_{Scrap} = \sum_{j=1}^{N_{Rep}+1} \left[SV \times \left(\frac{1}{(1+i)} \right)^{\left(T \times \frac{j}{N_{Rep}+1}\right)} \right] \tag{2.29}$$

where

- SV is the scrap value of the project components.
- i is the discount rate.
- T is the total lifespan of the project in years.
- N_{Rep} is the number of replacement cycles during the project's lifespan.
- j represents each replacement cycle.

2.7 Objective function

The objective function is designed to minimize both the COE and the LPSP. The COE is directly influenced by the system's LCC, which encompasses several components:

- Capital Costs $(C_{Initial\_Capital})$: Initial expenses for system installation.
- Operation and Maintenance Costs $(C_{O\&M})$: Ongoing costs for system upkeep.
- Replacement Costs (C_{Rep}): Costs for replacing system components over time.
- Fuel Costs (C_{Fuel}): Expenses associated with diesel generators.

These cost components are functions of key optimization variables such as the number of wind turbines (N_{WT}), number of PV panels (N_{PV}), number of battery storage units (N_{Bat}), and number of diesel generators (N_{DG}). Effective optimization of these variables is essential for minimizing both the COE and LPSP throughout the project's lifecycle.

$$\min(\text{COE} + \text{LPSP}) \tag{2.30}$$

Subject to:

The wind, solar PV energy resources, and the battery bank are subject to the following constraints:

$$N_{WT\_min} \leq N_{WT} \leq N_{WT\_max} \tag{2.31}$$

$$N_{PV\_min} \leq N_{PV} \leq N_{PV\_max} \tag{2.32}$$

$$N_{Bat\_min} \leq N_{Bat} \leq N_{Bat\_max} \tag{2.33}$$

where N_{WT} is the number of wind turbines, N_{PV} is the number of PV panels, and N_{Bat} is the number of battery cells.

At any given time (t), the energy stored in the battery bank is subject to the following constraints:

$$E_{Bat\_min} \leq E_{Bat}(t) \leq E_{Bat\_max} \tag{2.34}$$

where $(E_{Bat_{max}})$ and $(E_{Bat_{min}})$ represent the maximum and minimum energy storage levels of the battery bank, respectively. These values can be calculated as

$$E_{Bat_{max}} = \left(\frac{N_{Bat} \times V_{Bat} \times K_{Bat}}{1,000} \right) \times SOC_{Bat_{max}} \tag{2.35}$$

$$E_{Bat_{min}} = \left(\frac{N_{Bat} \times V_{Bat} \times K_{Bat}}{1,000} \right) \times SOC_{Bat_{min}} \tag{2.36}$$

where

- N_{Bat} is the number of battery cells.
- V_{Bat} is the battery voltage.
- K_{Bat} is the rated capacity of the battery (in Ah).
- $SOC_{Bat_{max}}$ is the maximum state of charge.
- $SOC_{Bat_{min}}$ is the minimum state of charge.

The states of charge (SOC) can be calculated as

$$[SOC_{Bat_{min}} = 1 - DOD]$$

$$[SOC_{Bat_{max}} = SOC_{Bat_{min}} - DOD]$$

where *DOD* is the depth of discharge of the battery.

Figure 2.5 illustrates the typical relationship between the life cycle of a lead-acid battery and its DOD. This relationship helps in understanding how the battery's longevity is affected by varying depths of discharge, emphasizing the trade-off between battery capacity and lifespan.

The DG operates with increased efficiency when it is subjected to higher loads. For this reason, DG's operational efficiency is optimized at loads up to 25% of its rated capacity, as used in the calculations of this study. Consequently, the DG will

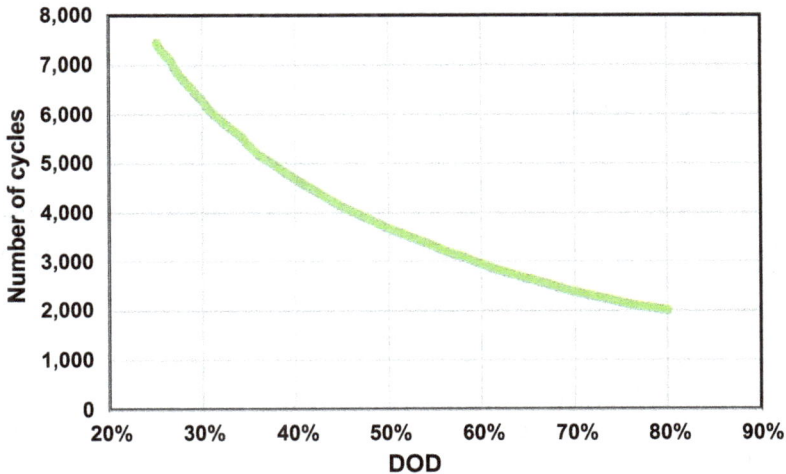

Figure 2.5 Typical lifecycle of RS lead-acid SSIG batteries as a function of DOD. This figure illustrates how the lifespan of RS lead-acid SSIG batteries varies with different levels of DOD. The graph demonstrates the trade-off between the battery's depth of discharge and its operational lifespan, highlighting the impact of DOD on battery longevity.

be activated only if the following conditions are met:

$$\frac{P_{Load}(t)}{\eta_{Conv}} \leq 0.25 \times P_{DG_{rating}} \times \Delta t \qquad (2.37)$$

where $P_{Load}(t)$ represents the load demand at each hour, η_{Conv} denotes the efficiency of the converter, $P_{DG_{rating}}$ is the rated power of the DG, and Δt refers to the simulation time duration.

LPSP and dump energy $\left(\left(E_{\{Dump\}}\right)\right)$ are critical indicators of system reliability. The optimization approach aims to minimize LPSP and dump load values to adhere to specific limits and achieve the lowest levelized energy cost. In this study, the acceptable limits for LPSP and dump energy are defined as follows:

$$LPSP \leq 5\% \text{ of } ADL \qquad (2.38)$$

$$E_{Dump} \leq 2\% \text{ of } ADL \qquad (2.39)$$

where ADL represents the total annual load demand, which can be calculated using

$$ADL = \sum_{t=1}^{8760} P_{Load}(t) \qquad (2.40)$$

Here, $P_{Load}(t)$ denotes the load demand at each hour.

2.8 Energy balance

1. Energy balance constraint:
    The total energy supplied by wind turbines $(P_{WT}(t))$, solar PV panels $(P_{PV}(t))$, the battery bank $(P_{Bat}(t))$, and the DG $\left(P_{\{DG\}(t)}\right)$ must meet or exceed the energy demand $(P_{Load}(t))$ at all times

$$P_{WT}(t) + P_{PV}(t) + P_{Bat}(t) + P_{DG}(t) \geq P_{Load}(t) \qquad (2.41)$$

where $P_{Bat}(t)$ is the power supplied by the battery bank, which can be positive (discharging) or negative (charging), and $P_{DG}(t)$ is the power generated by the DG.

2. Battery storage constraints:
    - Battery Charge Limit: The battery state of charge $\left(\left(E_{\{Bat\}(t)}\right)\right)$ must be within its minimum and maximum limits:

$$E_{Bat_{min}} \leq E_{Bat}(t) \leq E_{Bat_{max}} \qquad (2.42)$$

- Battery Charge and Discharge Dynamics: The battery's charging and discharging rates should adhere to the efficiency and maximum charge/discharge limits:

$$\text{Charging Rate} \leq \frac{E_{Bat}(t) - E_{Bat}(t-1)}{\Delta t} \leq \text{Discharging Rate} \qquad (2.43)$$

where (Δt) is the time step.

3. Renewable resource availability:
   - Wind Speed Constraints: The wind speed at hub height ($u(h)$) must be within the turbine's operational range, i.e., between cut-in speed (u_{cut-in}) and cut-off speed $\left(u_{cut-off} \right)$:

$$u_{cut-in} \leq u(h) \leq u_{cut-off} \qquad (2.44)$$

   - Solar Irradiance Constraints: The solar irradiance ($G(t)$) should be within the range of values that the PV panels can effectively convert to electricity.

4. Power conversion and storage constraints:
   - Bidirectional Converter Capacity: The power rating of the bidirectional converter ($P_{BDC-CC}(t)$) must be sufficient to handle the energy flows between the energy sources and the battery:

$$P_{BDC-CC} \geq \max(P_{WT}(t) - P_{Load}(t), P_{PV}(t) - P_{Load}(t), P_{Load}(t) - P_{WT}(t) - P_{PV}(t)) \qquad (2.45)$$

5. DG constraints:
   - Power Output Constraint: The power output of the DG ($P_{DG}(t)$) should be sufficient to cover any shortfall in energy supply:

$$P_{DG}(t) \geq \max(0, P_{Load}(t) - (P_{WT}(t) + P_{PV}(t) + P_{Bat}(t))) \qquad (2.46)$$

   - Fuel Consumption Constraint: The fuel consumption of the DG ($F_{DG}(t)$) should align with its rated consumption parameters and be accounted for in the economic evaluation.

6. Operational and maintenance constraints::
   - System Reliability: All components, including the DG, must operate within their specified reliability thresholds to ensure consistent performance and minimize downtime.
   - Maintenance Scheduling: Regular maintenance activities must be scheduled for the DG and other components to ensure that they do not disrupt the power supply.

These constraints ensure that the hybrid energy system, including the DG, operates efficiently, reliably, and within the technical and economic limits of its components. MATLAB Code 2.7 presents the implementation of the cost calculation.

```matlab
% Define parameters
i = 0.05; % Real net interest rate (5%)
T = 25; % Project lifetimes in years
P_Fuel = 0.80; % Price of fuel per liter in USD
SOC_Bat_max = 0.9; % Maximum State of Charge
SOC_Bat_min = 0.2; % Minimum State of Charge
DOD = SOC_Bat_max - SOC_Bat_min; % Depth of Discharge
P_DG_rating = 1000; % Rated power of diesel generator in
kW
P_Load = 0.5; % Load demand in kW
N_WT = 5; % Number of wind turbines
N_PV = 20; % Number of PV panels
N_Bat = 30; % Number of battery cells
N_DG = 1; % Number of diesel generators
C_WT = 2000; % Cost per kW of wind turbine
C_PV = 1500; % Cost per kW of PV panels
C_Bat = 500; % Cost per kWh of battery storage
C_DG = 5000; % Cost of diesel generator
C_BDC_CC = 3000; % Cost of bidirectional converter with
charge controller
V_Bat = 12; % Battery voltage
K_Bat = 100; % Battery rated capacity in Ah

% Initial Capital Cost Calculation
C_Initial_Capital = C_WT * N_WT + C_PV * N_PV + C_Bat *
N_Bat + C_DG * N_DG + C_BDC_CC; % Initial capital cost
disp(['Initial Capital Cost: $', num2str(C_Initial_Capital)])

% Operating and Maintenance Costs Calculation
C_O_M = C_Initial_Capital * 0.02; % Assuming 2% of initial costs
disp(['Operating and Maintenance Costs per year: $',
num2str(C_O_M)])

% Replacement Costs Calculation
N_Rep = 2; % Number of replacements anticipated
K_Rep = 1000; % Capacity of replacement components in kW
C_u = C_WT; % Cost of replacement components in $/kW
C_Rep = (K_Rep * C_u * ((1 + i)^T - 1) / (i * (1 +
i)^(N_Rep))) * 1e-3; % Present value of replacement costs
disp(['Replacement Costs: $', num2str(C_Rep)])

% Fuel Cost Calculation
F_DG = 0.2; % Fuel consumption in liters per kWh
C_Fuel = F_DG * P_Fuel; % Fuel cost
disp(['Fuel Cost: $', num2str(C_Fuel)])
```

MATLAB Code 2.7   MATLAB code for implementation of the cost calculation

```
% Scrap Value Calculation
V_Scrap = 0.1 * C_Initial_Capital; % Assuming salvage
value is 10% of initial capital cost
disp(['Scrap Value: $', num2str(V_Scrap)])

% Life Cycle Cost Calculation
LCC = C_Initial_Capital + (C_O_M * T) + C_Rep + C_Fuel -
V_Scrap;
disp(['Life Cycle Cost (LCC): $', num2str(LCC)])

% Capital Recovery Factor Calculation
CRF = i * (1 + i)^T / ((1 + i)^T - 1);
disp(['Capital Recovery Factor (CRF): ', num2str(CRF)])

% Cost of Energy Calculation
COE = (LCC * CRF) / (P_Load * 8760);
disp(['Cost of Energy (COE): $', num2str(COE)])

% Constraints (Sample, needs to be defined based on
actual system requirements)
E_Bat_max = (N_Bat * V_Bat * K_Bat / 1000) * SOC_Bat_max;
% Maximum energy storage
E_Bat_min = (N_Bat * V_Bat * K_Bat / 1000) * SOC_Bat_min;
% Minimum energy storage

disp(['Battery Energy Storage Limits: Min = ',
num2str(E_Bat_min), ' kWh, Max = ', num2str(E_Bat_max), '
kWh'])

% Assuming some constraints for Diesel Generator
Efficiency = 0.85; % Efficiency of the diesel generator
P_Load_max = 0.25 * P_DG_rating; % Max load for DG
operation
disp(['Diesel Generator Efficiency and Constraints:
Efficiency = ', num2str(Efficiency), ', Max Load = ',
num2str(P_Load_max)])

% Display all results
fprintf('Economic Analysis Results:\n')
fprintf('Initial Capital Cost: $%.2f\n',
C_Initial_Capital)
fprintf('Operation and Maintenance Costs per year:
$%.2f\n', C_O_M)
fprintf('Replacement Costs: $%.2f\n', C_Rep)
fprintf('Fuel Cost: $%.2f\n', C_Fuel)
```

MATLAB Code 2.7   (Continued)

```
fprintf('Scrap Value: $%.2f\n', V_Scrap)
fprintf('Life Cycle Cost (LCC): $%.2f\n', LCC)
fprintf('Cost of Energy (COE): $%.2f\n', COE)
fprintf('Battery Energy Storage Limits: Min = %.2f kWh,
Max = %.2f kWh\n', E_Bat_min, E_Bat_max)
fprintf('Diesel Generator Efficiency and Constraints:
Efficiency = %.2f, Max Load = %.2f kW\n', Efficiency,
P_Load_max)
```

MATLAB Code 2.7   (Continued)

2.9 Simulation results and discussion

In this section, two scenarios have been considered. The first scenario is without any method of DSM, while the second scenario is with a simple DSM. The pseudocode of the DSM method can be found in algorithm 1. The application of the SSA to determine the optimum configuration of the hybrid system has been conducted. The variables to determine are the WT, PV, DG, and battery. The objective function is to minimize the COE and LPSP. The convergence curve of the objective function is shown in Figure 2.6. The results of the optimum configuration are shown in Table 2.1.

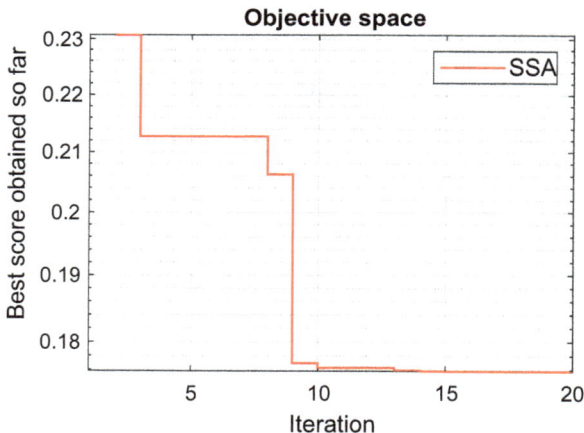

Figure 2.6.   Convergence curve of the SSA to minimize the cost function to optimally determine the RES configuration

Table 2.1   Optimum configuration of the RES

Variable	Value
n_PV	1,100
n_WT	14
n_DG	1.894405
n_batt	386.8292

Algorithm 1 Energy management system with load shifting—pseudocode summary

---

Initialize Parameters:
Set the simulation to run for a full year with hourly data.
Define profiles for load demand, wind power generation, solar power generation, and initial excess renewable energy.
Set inverter efficiency and daily time boundaries.

Energy Management Loop (for each hour in the year):
Determine Daily Boundaries for load shifting within the current day.

Handle Surplus Renewable Energy:
Check if there is surplus energy at the current hour.
Identify potential future hours within the same day with high load demand.
Shift load to the current hour if surplus energy can cover it, ensuring that any shifted load does not exceed 20% of the original future load.

Handle Renewable Energy Deficit:
If there is insufficient renewable energy to meet the current load demand, calculate the load deficit.
Identify future hours within the same day that may have excess renewable generation.
Shift current load to those future hours if they have enough surplus renewable energy, again limiting the shift to 20% of the current load.

Plotting and Visualizing the Results:
Create plots to visualize the adjusted load profile, renewable generation profiles (for wind and solar), and any remaining excess renewable energy over time.

Case 1:
In this case study, the real demand is analyzed without applying any DSM methods. The figures provide insights into the MG performance under these conditions. Additionally, Table 2.2 presents key performance indices of the MG, highlighting its operational characteristics and efficiency without DSM interventions.

Table 2.2   Performance of the optimized configuration of the RES

Parameter	Value
COE	0.198786014
LPSP	1.3638E−06
Dummy load ratio	0.003917281
Annual PV cost	15,217.34436
Annual WT cost	139,483.7929
Annual battery cost	22,021.66581
Annual DG cost	278,755.0365

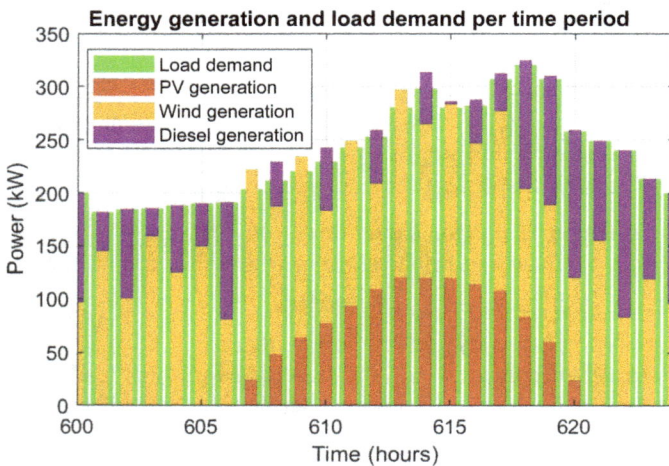

Figure 2.7   Energy generation from PV, wind, and diesel sources in comparison to load demand over time with the application of DSM

Figure 2.7 visualizes the alignment between real-time load demand and energy generation from PV, wind, and diesel sources. In this figure, the demand for load is represented by a wider bar for clarity, while PV, wind, and diesel generation are shown as a stacked bar chart. This layout helps illustrate how the different energy sources combine to meet demand at various times, particularly highlighting instances where the DG supports peak loads.

Figure 2.8 shows the average monthly diesel generation, capturing seasonal fluctuations in diesel dependency. The plot reveals how diesel generation increases during specific months to compensate for lower renewable energy generation, offering insight into the DG's essential role in ensuring energy availability across different seasons.

Figure 2.9 illustrates the average monthly fuel consumption of the DG, emphasizing the seasonal patterns in fuel use and associated cost implications.

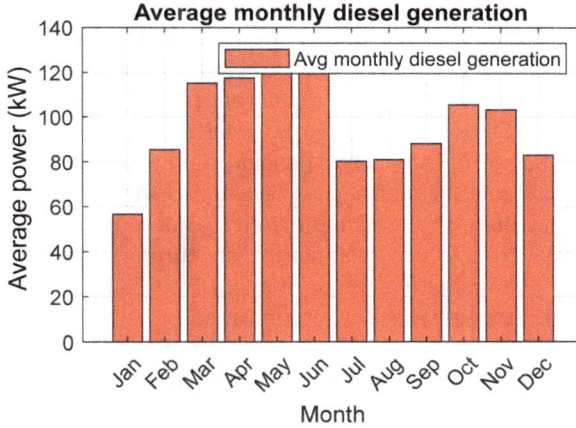

Figure 2.8   Average monthly diesel generation output, indicating diesel reliance during varying seasonal energy demand and renewable availability without the application of DSM

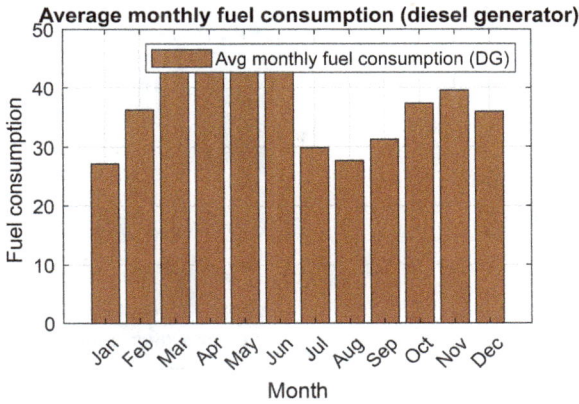

Figure 2.9   Average monthly fuel consumption for the DG, reflecting seasonal variations in fuel needs and cost implications without the application of DSM

Higher fuel consumption is observed in months with reduced renewable generation, indicating an increased reliance on diesel to meet demand. This trend underscores the cost impact of seasonal variability on fuel usage.

Figure 2.10 compares battery energy management for charging (E_CH) and discharging (E_DIS) on a monthly average basis, providing insight into the battery's cycling behavior. This subplot highlights the battery's role in energy storage during times of surplus generation, as well as its function in meeting shortfalls when demand exceeds renewable generation, with clear seasonal variation.

Figure 2.11 illustrates the battery state of charge (SOC) over a selected period, showing how SOC fluctuates in response to changes in load demand and renewable energy availability. The charge–discharge cycle represented here highlights the battery's contribution to dynamically balancing supply and demand.

Figure 2.12 presents the annual cost breakdown for each system component, including PV, wind, diesel, battery, and inverter. This comparison provides a clear view of the relative costs of each component, supporting an economic assessment of the system. It highlights the components that contribute most to annual costs, informing potential strategies for cost optimization.

Figure 2.13 visualizes the LPSP over time, indicating periods when energy supply falls short of demand. Tracking LPSP is crucial for evaluating the reliability of the system, especially during hours of high demand, and helps identify intervals where system performance could be improved.

Figure 2.14 shows the DG operational units over time, illustrating how the number of operational units adjusts in response to demand when renewable generation is low. This stair plot provides insight into the DG's role in maintaining

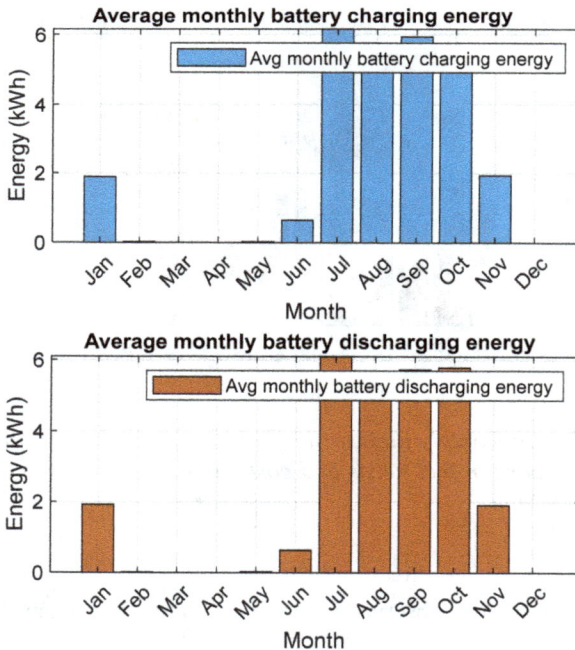

Figure 2.10   Average monthly energy for battery charging and discharging, showing seasonal battery utilization patterns without the application of DSM

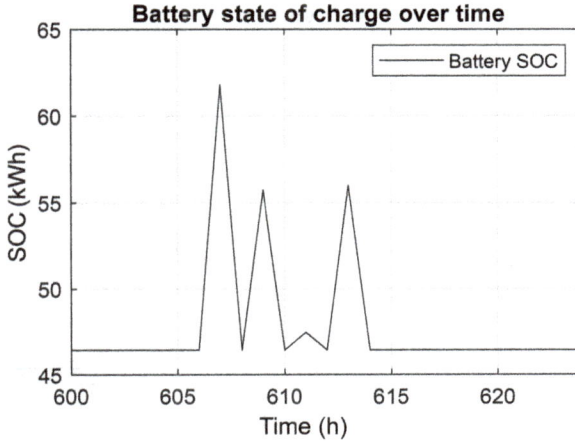

Figure 2.11   Battery SOC over time, highlighting charge–discharge cycles in response to load and generation dynamics without the application of DSM

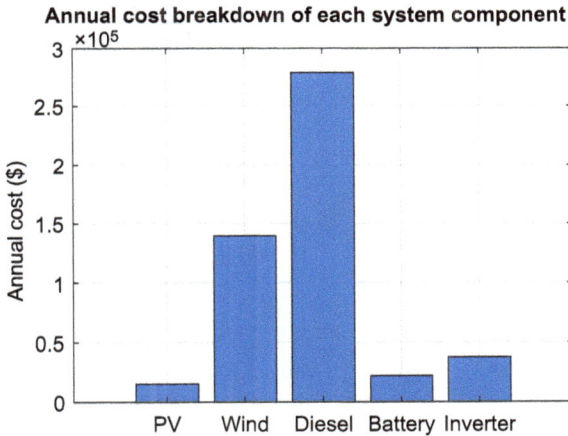

Figure 2.12   Annual cost breakdown by system component, facilitating economic analysis of each component's contribution to overall costs without the application of DSM

system reliability and stability, particularly during periods when renewable sources are insufficient to meet demand.

Case 2:

In this case study, the load demand is evaluated after applying DSM strategies. Figure 2.15 illustrates the adjusted load profile following DSM implementation,

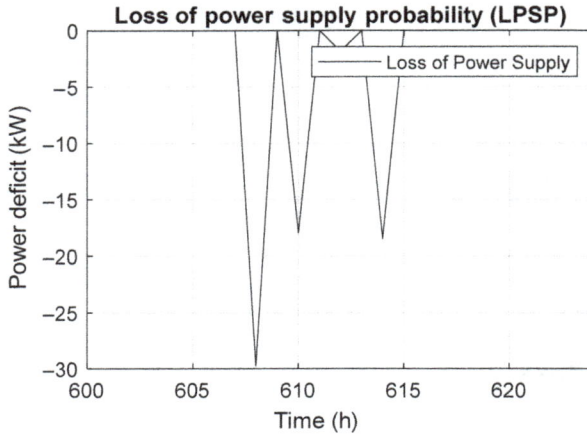

Figure 2.13   LPSP over time, depicting periods of power deficit relative to demand without the application of DSM

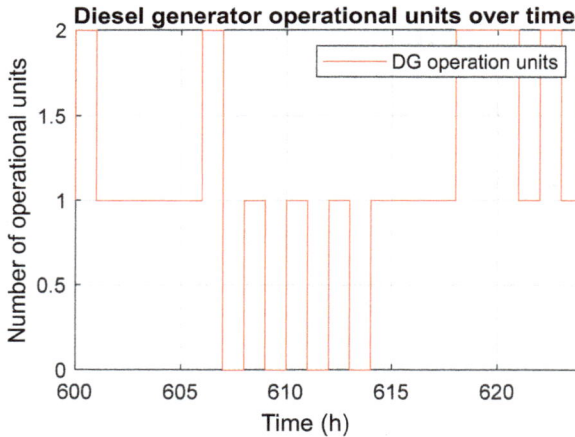

Figure 2.14   DG operational units over time, reflecting system response to demand changes and renewable variability without the application of DSM

providing insights into its impact on the MG performance. Additionally, the table presents key performance indices for the MG, highlighting improvements in operational efficiency and load balancing achieved through DSM.

Energy generation and load demand per time period is analyzed in Figure 2.16. This figure illustrates the temporal alignment of load demand with energy generation from PV, wind, and diesel sources. The load demand is

Energy management system results

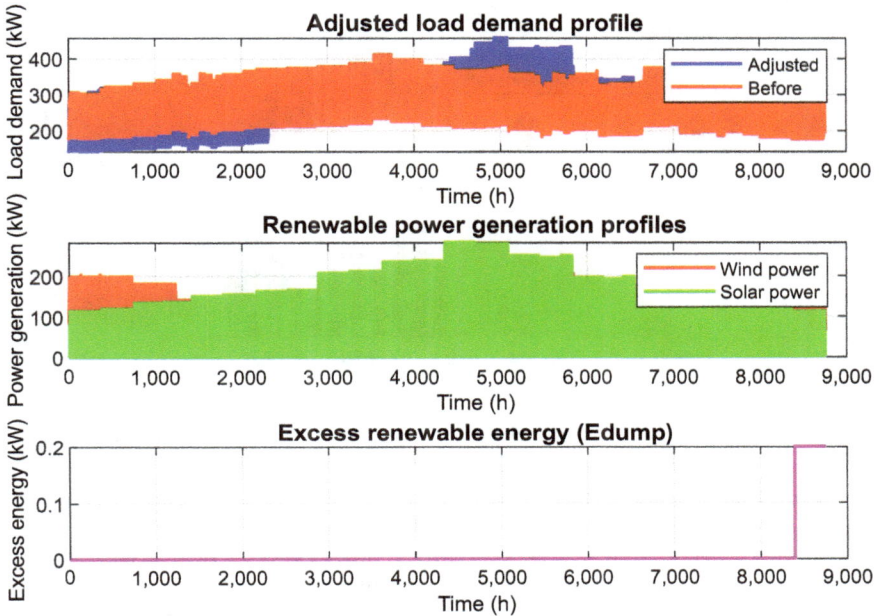

Figure 2.15 The adjusted load profile following DSM implementation

Table 2.3 The MG performance indices after applying the DSM

Parameter	With DSM
COE	0.196869
LPSP	0.014826
Dummy load ratio	0
Annual PV cost	15217.34
Annual WT cost	139483.8
Annual battery cost	22021.67
Annual DG cost	274363.7

represented as a wider bar for clear comparison, while PV, wind, and diesel generation are displayed as a stacked bar chart. This layout helps visualize how different energy sources contribute to meeting the demand, particularly during peak and off-peak periods.

Moreover, the average monthly diesel generation is shown in Figure 2.17. The bar plot shows the average monthly DG output, revealing seasonal trends. Diesel generation

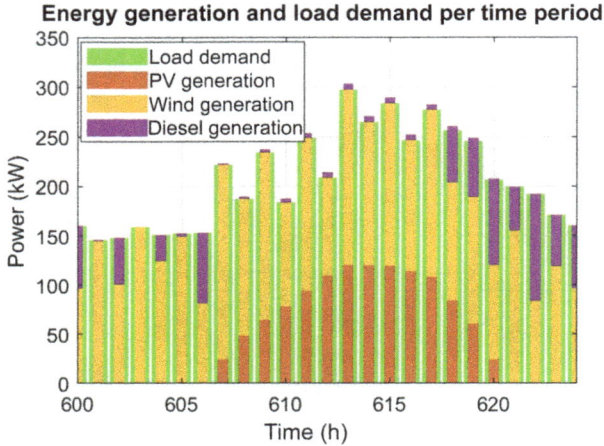

Figure 2.16   Energy generation from PV, wind, and diesel sources in comparison to load demand over time with the application of DSM

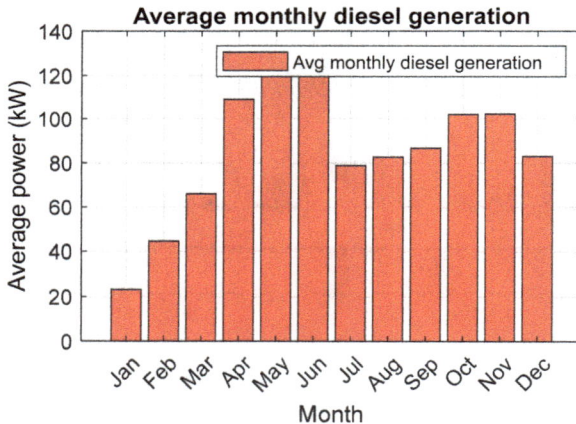

Figure 2.17   Average monthly diesel generation output, indicating diesel reliance during varying seasonal energy demand and renewable availability with the application of DSM

increases in certain months when renewable sources are insufficient to meet demand, providing insight into diesel dependency fluctuations throughout the year.

The average monthly fuel consumption of the DG is shown in Figure 2.18. This figure presents the monthly fuel consumption for the DG, emphasizing fuel dependency and cost implications associated with seasonal diesel usage. It

Average monthly fuel consumption (diesel generator)

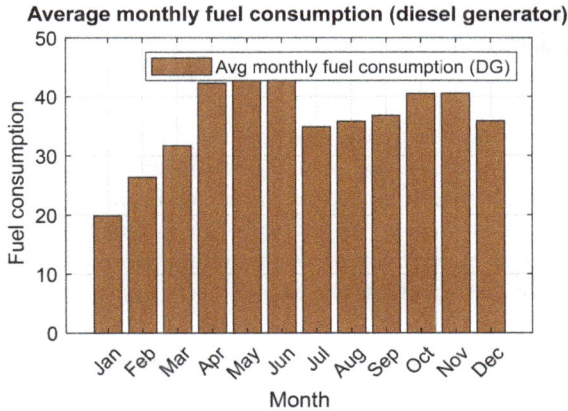

Figure 2.18   Average monthly fuel consumption for the DG, reflecting seasonal variations in fuel needs and cost implications with the application of DSM

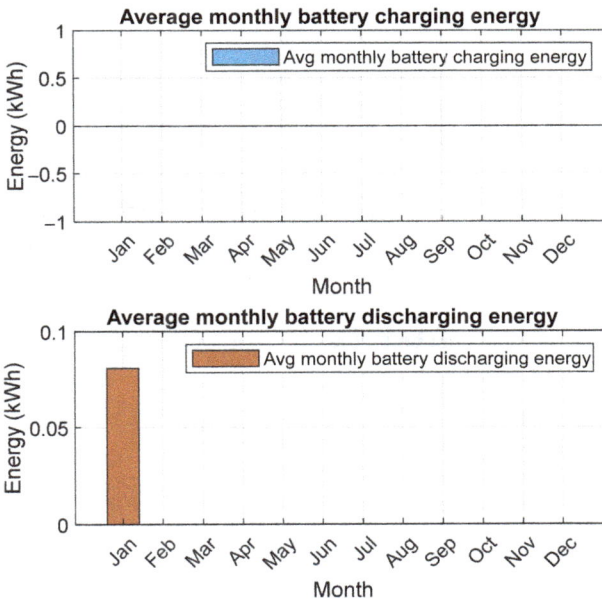

Figure 2.19   Average monthly energy for battery charging and discharging, showing seasonal battery utilization patterns with the application of DSM

highlights months with higher fuel consumption, often correlating with low renewable generation periods.

Figure 2.19 shows the battery energy management for the charging and discharging status. This subplot compares the monthly average energy for battery charging (E_CH) and discharging (E_DIS), providing insight into battery cycling patterns. The distinction between charging and discharging illustrates the battery's role in storing excess energy and meeting shortfalls, with seasonal variation in activity.

The battery state of charge over time is shown in Figure 2.20. The plot shows the battery's SOC over a selected time frame, illustrating the battery's charge–discharge cycle in response to load fluctuations and renewable generation availability. SOC trends help evaluate the battery's role in maintaining energy balance.

The annual cost breakdown of each system component is displayed in Figure 2.21. The figure represents the annual costs associated with each system

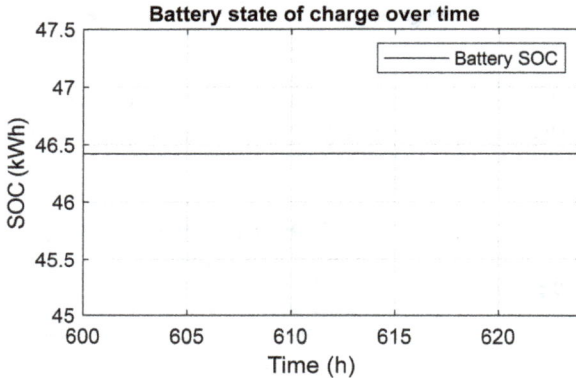

Figure 2.20   Battery SOC over time, highlighting charge–discharge cycles in response to load and generation dynamics with the application of DSM

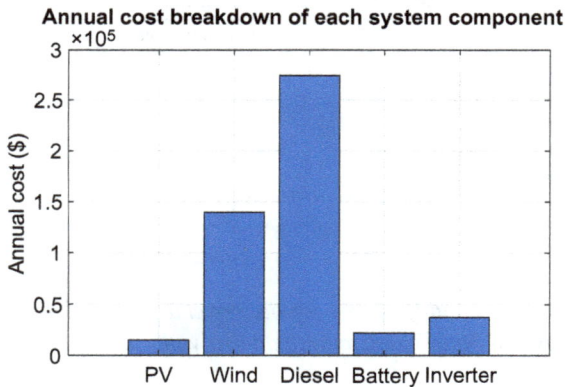

Figure 2.21   Annual cost breakdown by system component, facilitating economic analysis of each component's contribution to overall costs with the application of DSM

component (PV, wind, diesel, battery, and inverter), aiding in a comparative eco-nomic assessment. It shows which components contribute most to the system's annual costs, informing optimization strategies.

Figure 2.22 visualizes the LPSP. The figure illustrates the probability of power deficits (LPSP) over time, indicating periods when energy supply fails to meet demand. Monitoring LPSP is crucial for assessing system reliability, particularly during high-demand hours.

The DG operational units over time are shown in Figure 2.23. The stair plot visualizes the number of operational DG units at different time intervals, demon-strating how the system scales diesel generation in response to demand or

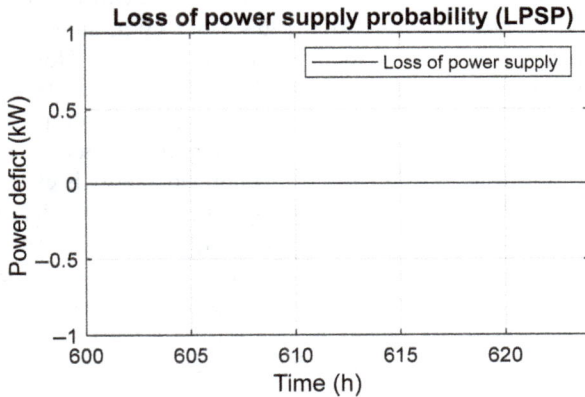

Figure 2.22   *LPSP over time, depicting periods of power deficit relative to demand with the application of DSM*

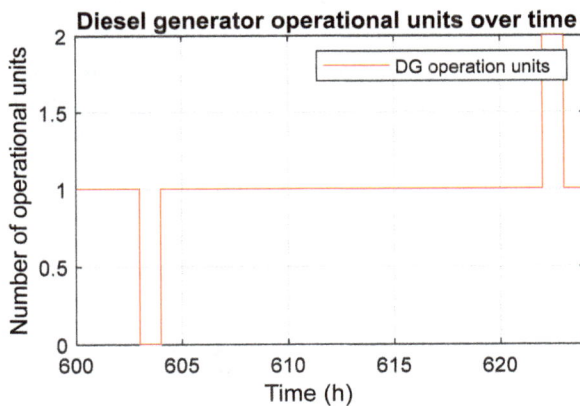

Figure 2.23   *DG operational units over time, reflecting system response to demand changes and renewable variability with the application of DSM*

renewable generation shortfalls. It helps assess the DG's role in providing relia-bility and stability.

2.10   Discussion

The comparison of system performance with and without DSM reveals notable differences in key parameters, highlighting the benefits of implementing DSM in optimizing hybrid energy systems. The data in Table 2.4 highlights sig-nificant differences in system performance, such as COE, LPSP, and dummy load ratio, when DSM is applied. The data presented in the table compares the system performance and economic parameters with and without DSM. The COE decreases slightly with DSM, from 0.1988 to 0.1969 USD/kWh, indicating a marginal reduction in energy costs due to optimized load shifting. However, the LPSP increases notably with DSM, rising from 1.36×10^{-6} to 0.0148, suggesting that DSM introduces a slight risk of unmet demand, although it remains within acceptable levels. The dummy load ratio, which represents unused energy, is reduced to zero with DSM, ensuring that all generated energy is efficiently utilized. The annual costs for PV, WT, and battery systems remain unchanged, while the annual DG cost decreases from 278,755.04 to 274,363.67 USD, reflecting reduced reliance on diesel power. Additionally, DSM significantly reduces the battery operation hours from 1267 to only 3 h, indicating a substantial decrease in battery use. The total charging energy for the battery drops to zero with DSM, while the total discharging energy is minimal, further emphasizing the reduced dependency on battery storage when DSM is applied.

In terms of system components, the number of photovoltaic modules (n_{PV}), wind turbines (n_{WT}), diesel generators (n_{DG}), and batteries (n_{batt}) remains unchan-ged regardless of the presence of DSM. This consistency suggests that DSM operates effectively within the existing infrastructure, utilizing the current system

Table 2.4   Comparison of key system performance parameters with and without DSM. The table shows the effect of DSM on the COE, LPSP, dummy load ratio, component costs, and battery operation hours.

Parameter	Without DSM	With DSM
COE	0.198786014	0.196869473
LPSP	1.3638E-06	0.014826107
Dummy load ratio	0.003917281	0
Annual PV cost	15,217.34436	15,217.34436
Annual WT cost	139,483.7929	139,483.7929
Annual Battery cost	2,2021.66581	22,021.66581
Annual DG cost	278,755.0365	274,363.6704
Operation hours (battery)	1,267	3
Total charging energy (kWh)	20,495.31061	0
Total discharging energy (kWh)	20,298.26272	58.93318186

capacity without necessitating any additional investments or adjustments to component sizes. This aspect underscores the ability of DSM to optimize system performance without requiring substantial modifications to the system setup.

The total load ($P_{\text{tot\_load}}$) is identical in both scenarios, which confirms that DSM is employed to manage demand rather than reduce the overall energy consumption. DSM likely functions by shifting load profiles to optimize the timing of energy usage, thus improving resource utilization without affecting the overall energy demand. This strategic load shifting enhances system performance by aligning energy consumption with the availability of renewable resources, ensuring that demand is met efficiently.

One of the key benefits of DSM is its impact on the COE. With DSM in place, the COE decreases slightly from 0.1988 to 0.1969 USD/kWh. This reduction reflects the cost-saving potential of DSM, which helps minimize reliance on more expensive energy sources or operations. By optimizing the timing and distribution of load, DSM can reduce the need for costly energy generation methods, thus improving the overall economic efficiency of the system.

The LPSP also shows an interesting change. Without DSM, the LPSP is extremely low (1.36×10^{-6}), but with DSM, it rises to 0.0148. While this increase suggests that DSM may shift some energy demand to periods when renewable generation is less available, the resulting LPSP remains acceptably low. This indicates that DSM still maintains a high level of system reliability, even with the slight increase in the likelihood of unmet demand. Thus, DSM proves to be a reliable strategy for balancing the load while offering substantial cost-saving benefits.

A significant impact of DSM is observed in the dummy load ratio, which represents the unused generated energy. With DSM, the dummy load ratio is reduced to zero, compared to approximately 0.39% without DSM. This improvement suggests that DSM plays a crucial role in ensuring that all generated energy is effectively utilized. By better aligning demand with production, DSM minimizes wasted energy, enhancing the system's overall energy efficiency.

The annual costs associated with the system components remain largely unchanged between the two scenarios, with the exception of the annual DG cost, which decreases from 278,755.04 to 274,363.67 USD with DSM. This reduction can be attributed to DSM's role in optimizing the use of renewable energy and minimizing reliance on diesel generators. By shifting load to times when renewable energy generation is more abundant, DSM reduces the runtime of the diesel generators, resulting in lower operational costs. This reduction in diesel generation costs highlights one of the key advantages of DSM: reducing dependence on fossil-fuel-based power generation and supporting the use of cleaner, renewable energy sources.

Overall, the implementation of DSM in the hybrid energy system leads to several benefits. It reduces the COE, leading to financial savings; improves energy efficiency by eliminating the dummy load; enhances the use of renewable energy, thereby reducing reliance on diesel generators; and maintains high system reliability with only a minor increase in the LPSP. In conclusion, DSM is a valuable strategy that enhances the sustainability, cost-effectiveness, and efficiency of hybrid energy systems, making it an important tool for optimizing energy consumption and reducing the environmental impact of energy generation.

2.11   Conclusion

In this chapter, the impact of DSM on the performance and economics of a hybrid renewable energy system was analyzed. The results highlighted that DSM optimizes energy usage and reduces costs by shifting demand to align with renewable energy generation, thus decreasing reliance on fossil fuel-based sources and lowering operational costs, especially for diesel generators. While DSM caused a slight increase in the LPSP, the value remained within acceptable limits, demonstrating that DSM maintains system reliability. Additionally, DSM eliminated the dummy load ratio, ensuring more efficient use of generated energy and improving overall system efficiency. The application of DSM did not require changes in system components, making it a cost-effective strategy for optimizing performance without significant infrastructure investment. Overall, DSM proves to be a valuable tool for enhancing both the economic viability and environmental sustainability of hybrid energy systems, supporting the transition to more efficient, renewable-powered operations. Moreover, the chapter provides a collection of valuable MATLAB codes that are integral to the implementation of the hybrid renewable energy system with DSM. These codes serve as practical tools for modeling, simulating, and optimizing the system, enabling the effective integration of DSM strategies into real-world applications. Future research could explore advanced DSM strategies, energy storage integration, and their broader impact on system performance, grid stability, and energy security.

References

[1]   T. P. Brennand, "Renewable energy in the United Kingdom: policies and prospects," *Energy for Sustainable Development*, vol. 8, no. 1, pp. 82–92, 2004, doi:10.1016/s0973-0826(08)60393-2.

[2]   "IEA Wind Energy Task 37 – System Engineering – Aerodynamic Optimization Case Study," American Institute of Aeronautics and Astronautics (AIAA), 2020.

[3]   A. A. Z. Diab, H. M. Sultan, and O. N. Kuznetsov, "Optimal sizing of hybrid solar/wind/hydroelectric pumped storage energy system in Egypt based on different meta-heuristic techniques," *Environmental Science and Pollution Research*, vol. 27, no. 26, pp. 32318–32340, 2019, doi:10.1007/s11356-019-06566-0.

[4]   S. Ullah, A. M. A. Haidar, P. Hoole, H. Zen, and T. Ahfock, "The current state of distributed renewable generation, challenges of interconnection and opportunities for energy conversion based DC microgrids," *Journal of Cleaner Production*, vol. 273, p. 122777, 2020, doi:10.1016/j.jclepro.2020.122777.

[5]   R. Medara and R. S. Singh, "Energy efficient and reliability aware workflow task scheduling in cloud environment," *Wireless Personal Communications*, vol. 119, pp. 1301–1320, 2021, doi:10.1007/s11277-021-08263-z.

[6]   H. Canton, "International Energy Agency—IEA," in *The Europa Directory of International Organizations 2021*, Routledge, 2021, pp. 684–686.

[7]   A. M. Eltamaly, M. A. Alotaibi, W. A. Elsheikh, A. I. Alolah, and M. A. Ahmed, "Novel Demand Side-Management Strategy for Smart Grid Concepts Applications in Hybrid Renewable Energy Systems," in *2022 4th International Youth Conference on Radio Electronics, Electrical and Power Engineering (REEPE)*, 2022, http://dx.doi.org/10.1109/reepe53907.2022. 9731431.

[8]   E. Fedele, D. Iannuzzi, and A. Del Pizzo, "Onboard energy storage in rail transport: Review of real applications and techno-economic assessments," *IET Electrical Systems in Transportation*, vol. 11, no. 4, pp. 279–309, 2021, doi:10.1049/els2.12026.

[9]   A. A. Z. Diab, A. M. El-Rifaie, M. M. Zaky, and M. A. Tolba, "Optimal sizing of stand-alone microgrids based on recent metaheuristic algorithms," *Mathematics*, vol. 10, no. 1, p. 140, 2022, doi:10.3390/ math10010140.

[10] K. Abedrabboh and L. Al-Fagih, "Applications of mechanism design in market-based demand-side management: A review," *Renewable and Sustainable Energy Reviews*, vol. 171, p. 113016, 2023, doi:10.1016/j.rser.2022. 113016.

[11] A. M. Eltamaly, "A novel energy storage and demand side management for entire green smart grid system for NEOM city in Saudi Arabia," *Energy Storage*, vol. 6, no. 1, p. e515, 2023, doi:10.1002/est2.515.

[12] A. M. Eltamaly, "Smart decentralized electric vehicle aggregators for optimal dispatch technologies," *Energies*, vol. 16, no. 24, p. 8112, 2023, doi:10. 3390/en16248112.

[13] R. D. Ganoe, P. W. Stackhouse Jr, and R. J. DeYoung, "RETScreen plus software tutorial," 2014.

[14] E. I. Baring-Gould, "Hybrid2: The hybrid system simulation model, Version 1.0, user manual," National Renewable Energy Lab.(NREL), Golden, CO (United States), 1996.

[15] A. A. Khan and A. F. Minai, "A strategic review: The role of commercially available tools for planning, modelling, optimization, and performance measurement of photovoltaic systems," *Energy Harvesting and Systems*, vol. 11, no. 1, p. 20220157, 2024.

[16] A. Rana and G. Gróf, "Assessment of prosumer-based energy system for rural areas by using TRNSYS software," *Cleaner Energy Systems*, vol. 8, p. 100110, 2024.

[17] H. Abubakr, J. C. Vasquez, K. Mahmoud, M. M. F. Darwish, and J. M. Guerrero, "Comprehensive review on renewable energy sources in Egypt— Current status, grid codes and future vision," *IEEE Access*, vol. 10, pp. 4081– 4101, 2022, doi:10.1109/access.2022.3140385.

[18] C. Breyer, S. Khalili, D. Bogdanov, *et al.*, "On the history and future of 100% renewable energy systems research," *IEEE Access*, vol. 10, pp. 78176–78218, 2022, doi:10.1109/access.2022.3193402.

[19] S. A. Al Dawsari, F. Anayi, and M. Packianather, "Novel Optimal Configuration Approach for off-grid microgrid with Hybrid Energy Storage Using Mantis Search Algorithm," presented at the 2024 12th International Conference on Smart Grid (icSmartGrid), 2024/05/27, 2024 [Online]. Available: http://dx.doi.org/10.1109/icsmartgrid61824.2024.10578170.

[20] I. Us Salam, M. Yousif, M. Numan, and M. Billah, "Addressing the challenge of climate change: The role of microgrids in fostering a sustainable future – A comprehensive review," *Renewable Energy Focus*, vol. 48, p. 100538, 2024, doi:10.1016/j.ref.2024.100538.

[21] E. E. E. Frost Sullivan, N. Townsend and C. Ahlfeldt, "Development of a Customised Sector Programme for Small Scale Renewable Energy in South Africa," in *Task 1-5 Final Report*, 2013. [Online]. Available: http://www.wasaproject.info/docs/dtiSmallScaleRenewablesFinalReport.pdf [Accessed 3/2/2025].

[22] M. Wang, X. Mao, Y. Xing, *et al.*, "Breaking down barriers on PV trade will facilitate global carbon mitigation," *Nature Communications*, vol. 12, no. 1, pp. 1–16, 2021.

[23] J. Lian, Y. Zhang, C. Ma, Y. Yang, and E. Chaima, "A review on recent sizing methodologies of hybrid renewable energy systems," *Energy Conversion and Management*, vol. 199, p. 112027, 2019.

[24] S. Islam, S. Mostaghim, and M. Hartmann, "A survey on multi-objective optimization in microgrid systems," 2024: IEEE, pp. 1–8.

[25] A. A. Rathod and B. Subramanian, "Scrutiny of hybrid renewable energy systems for control, power management, optimization and sizing: Challenges and future possibilities," *Sustainability*, vol. 14, no. 24, p. 16814, 2022.

[26] K. Alqunun, "Strength pareto evolutionary algorithm for the dynamic economic emission dispatch problem incorporating wind farms and energy storage systems," *Engineering, Technology & Applied Science Research*, vol. 10, no. 3, pp. 5668–5673, 2020.

[27] A. Heydari, M. M. Nezhad, F. Keynia, *et al.*, "A combined multi-objective intelligent optimization approach considering techno-economic and reliability factors for hybrid-renewable microgrid systems," *Journal of Cleaner Production*, vol. 383, p. 135249, 2023.

[28] O. Nadjemi, T. Nacer, A. Hamidat, and H. Salhi, "Optimal hybrid PV/wind energy system sizing: Application of cuckoo search algorithm for Algerian dairy farms," *Renewable and Sustainable Energy Reviews*, vol. 70, pp. 1352–1365, 2017.

[29] A. M. Ibrahim, M. A. Attia, M. M. Othman, and A. Y. Abdelaziz, "A survey on demand side management in electrical power systems," *I-Manager's Journal on Power Systems Engineering*, vol. 5, no. 1, 2017.

[30] A. T. Dahiru, D. Daud, C. W. Tan, Z. T. Jagun, S. Samsudin, and A. M. Dobi, "A comprehensive review of demand side management in distributed

grids based on real estate perspectives," *Environmental Science and Pollution Research*, vol. 30, no. 34, pp. 81984–82013, 2023.

[31] R. Debnath, D. Kumar, and D. K. Mohanta, "Effective demand side management (DSM) strategies for the deregulated market envioronments," 2017: IEEE, pp. 110–115.

[32] C. F. R. Augusto, "Evaluation of the potential of demand side management strategies in PV system in rural areas," PhD thesis, Universidade de Lisboa, 2017. [Online]. Available: https://repositorio.ulisboa.pt/handle/10451/30607 [Accessed 3/2/2025].

[33] P. Sanjeev, N. P. Padhy, and P. Agarwal, "Peak energy management using renewable integrated DC microgrid," *IEEE Transactions on Smart Grid*, vol. 9, no. 5, pp. 4906–4917, 2017.

[34] A. Molderink, V. Bakker, M. G. C. Bosman, J. L. Hurink, and G. J. M. Smit, "Management and control of domestic smart grid technology," *IEEE Transactions on Smart Grid*, vol. 1, no. 2, pp. 109–119, 2010.

[35] H. Balakrishnan, K. K. S. Tomar, and S. N. Singh, "An agent based approach for efficient energy management of microgrids," 2017: IEEE, pp. 1–5.

[36] S. Saini, "Demand-side management module," *Sustainable Energy Regulation and Policymaking for Africa*, vol. 14, no. 8, pp. 1–100, 2007.

[37] A. M. Samatar, S. Mekhilef, H. Mokhlis, M. Kermadi, and O. Alshammari, "Optimal design of a hybrid energy system considering techno-economic factors for off-grid electrification in remote areas," *Clean Technologies and Environmental Policy*, 2024, doi:10.1007/s10098-024-02939-3.

[38] H. H. H. Mousa, K. Mahmoud, and M. Lehtonen, "Recent developments of demand-side management towards flexible DER-rich power systems: A systematic review," *IET Generation, Transmission & Distribution*, vol. 18, no. 13, pp. 2259–2300, 2024, doi:10.1049/gtd2.13204.

Chapter 3

Addressing the economic dispatch solution of a smart microgrid with a developed charging strategy for electric vehicles with demand-side response

Mohamed Ebeed[1,2], Francisco Jurado[2] and Ali Selim[2,3]

Abstract

Reducing the operation cost and the dependence on fossil fuel-based energy resources is a crucial task in electric power systems, which can be accomplished via optimal economic dispatch (ED) problem solutions. In this regard, this work proposed a new modified dandelion optimizer (MDO) for solving the ED of a grid-connected microgrid, which consists of a diesel generator, three fuel cells (FCs), and two wind turbines to provide the required energy to electric loads and an electric vehicle (EV) charging station. The MDO is mainly based on two approaches, including the fitness distance balance and local elite movement to emphasize the searching ability of the conventional DO. In this chapter, a smart strategy is proposed for reducing the operation costs, which is based on the application of a demand-side response-based real-time price and a smart charging strategy for EVs, which is based on maximizing and controlling the charging powers of EVs at each hour instead of charging the EVs by uncontrolled power. The results reveal that the MDO is superior for ED solution compared to the grey wolf optimizer, whale optimization algorithm, sand cat swarm optimizers, and dandelion optimizer. In addition to that the cost is reduced considerably with the application of the smart strategy. Additionally, the operating cost is reduced considerably via the application of the DSR and the smart charging strategy.

Keywords: Economic dispatch; Demand-side response; Electric vehicles; Smart microgrid; Smart charging

[1]Department of Electrical Engineering, Faculty of Engineering, Sohag University, Egypt
[2]Department of Electrical Engineering, University of Jaén, Spain
[3]Department of Electrical Engineering, Faculty of Engineering, Aswan University, Egypt

3.1 Introduction

Recently, the use of electric vehicles (EVs) has increased massively, which is 17 million EVs in transportation sectors in 2024. The EVs' sales increased by 25% in the first quarter of 2024 compared to the same quarter in 2023. The expected growths of the EV market are 45%, 25%, and 11% in China, Europe, and the United States, respectively [1].

The transition to sustainable energy systems is progressively critical in addressing global climate change and reducing greenhouse gas emissions [2]. As a result, smart microgrids are a good way to combat global climate change. Unlike conventional grids, which depend on centralized power generation and are susceptible to widespread outages, smart microgrids use real-time data to optimize energy use, generate power locally, and continue to function even in the event of a failure. Additionally, smart microgrids prioritize the integration of renewable energy sources, which lowers emissions and gives customers more control over their energy use for increased economy and efficiency [3]. Recently, the increasing popularity of EVs presents both possibilities and distinct obstacles for maximizing energy dispatch when incorporating them into smart microgrids [4].

The integration of a smart charging strategy for EVs into the microgrid framework can significantly influence operational costs [5]. The demand side response based real time pricing (DSR-RTP) in which the electricity prices fluctuate throughout the day [6]. This approach may be computed using economic dispatch (ED) and results in further cost savings by lowering the grid's peak demand and optimizing the usage of renewable energy.

Hence, ED refers to the process of determining the optimal generation levels of various power sources in a way that minimizes costs while meeting load demands [7]. The importance of optimizing ED in microgrids is underscored by the diverse energy sources they incorporate renewable energies, such as wind and solar, as well as conventional fossil fuels [8].

The ED problem can now be solved more successfully because of recent developments in optimization methods [7]. To address ED, conventional techniques have been presented, and some of these techniques include dynamic programming [9], mixed integer linear [10], nonlinear [11], and linear approaches [12]. Nevertheless, conventional methods frequently fail to meet the complexity of contemporary microgrid devices. In response, researchers have turned to nature-inspired optimization algorithms [13], which emulate biological processes to find optimal solutions.

Metaheuristics are optimization methods that employ effective search techniques to explore large solution spaces and identify optimal solutions. A variety of algorithms, such as those based on human behavior, swarm intelligence techniques, evolutionary algorithms, and nature-inspired techniques, methodically improve potential solutions through repeated processes and frequently provide efficient, nearly optimal results. The ED problem has been solved using a variety of metaheuristic optimization techniques, including genetic algorithms (GA) [14], particle swarm optimization (PSO) [15], ant colony optimization (ACO) [16], and artificial bee colony (ABC) algorithms [17].

Metaheuristic algorithms have recently advanced to the point that researchers are using these methods to solve ED problems. In order to account for transmission power losses, valve-point loading effects, and restricted operating zones, the manta ray foraging optimization method has been utilized in the previous study [18]. To address the complex real-world issues of dynamic ED, the enhanced Cheetah optimizer algorithm was created [19]. The issue of environmental ED has been addressed by the proposal of an enhanced quantum particle swarm optimization algorithm [20]. A solution to the multi-area ED problem has been proposed: an enhanced competitive swarm optimization [21]. Nevertheless, despite their advantages, metaheuristic algorithms have drawbacks as well. For example, they cannot provide a global optimum in extremely nonlinear search spaces, and they need careful parameter adjustment, which may result in results that are suboptimal and sensitive to beginning circumstances.

Dandelion optimizer (DO) is a powerful and new optimization method that simulates the movements of the dandelion's seeds [22]. The DO is applied for solving numerous problems like load frequency control [23,24], maximizing the PV output power units [24], parameter identification of PV modules [25], revolutionizing photovoltaic panels [26], load frequency control of a multi-area system [27], and optimal setting of model predictive control for autonomous vehicles [28].

Among these, DO has shown promise and solved different problems, but it failed to solve the highly nonlinear and convex problem, and it is trapped in local optima, but its potential can be further enhanced through modifications, as proposed in this study by introducing the modified Dandelion optimizer (MDO). The contributions are outlined as follows:

- Solving the ED of a grid-tied hybrid MG with conventional and renewable energy resources for an electric load and EV charging station.
- A new MDO is proposed with the integration of fitness distance balance (FDB) and a local elite movement.
- The ED of the hybrid MG is solved with the application DSR and the smart charging strategy of EVs.
- The results obtained by the MDO are validated via a comparison with other optimization methods like grey wolf optimizer (GWO), whale optimization algorithm (WOA), sand cat swarm optimizers (SCSO), and the standard dandelion optimizer (DO).

The structure of the chapter is organized as follows: Section 3.2 outlines the mathematical formulation of the ED problem. Section 3.3 details the key steps of the DO algorithm. Section 3.4 introduces the modifications to the DO algorithm and its application to solve the ED problem. Section 3.5 presents the obtained results, and finally, Section 3.6 provides the conclusion.

3.2 Problem formulation

The aim of the ED of the isolated MG is to decrease the operating cost during the day ahead. The construction of the studied MG is shown in Figure 3.1, which

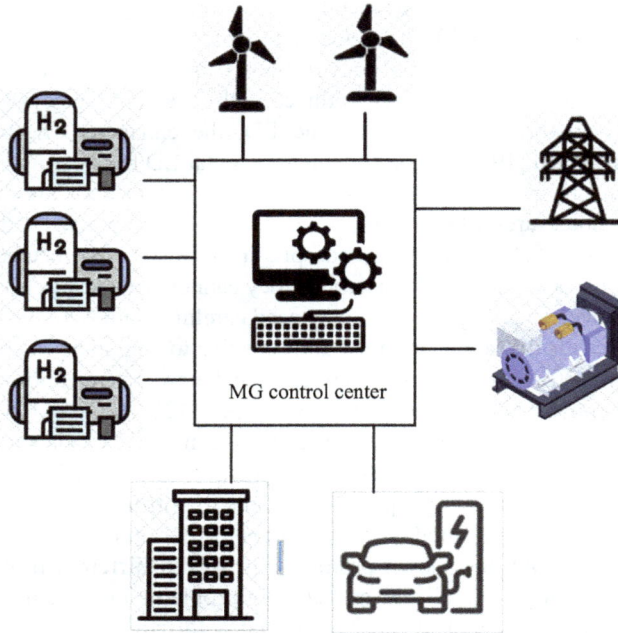

Figure 3.1   The configuration of the MG

consists of three wind turbines (WTs), one diesel generator, and three fuel cells (FCs). The objective function is to reduce the operating cost of the MG (Cost) as follows [29]:

$$\text{Cost} = \text{Cost}^{\text{WT}} + \text{Cost}^{\text{DG}} + \text{Cost}^{\text{FC}} \tag{3.1}$$

where Cost^{WT}, Cost^{DG}, and Cost^{FC} are operating costs of WTs, DGs, and FCs, respectively. The cost of WTs can be obtained using (3.2):

$$\text{Cost}^{\text{WT}} = \sum_{n=1}^{\text{NWT}} \left(C_n P_{W,n} \right) \tag{3.2}$$

in which

$$P_W(v) = \begin{cases} 0 & \text{for} \quad v < v_i \text{ and } v > v_o \\ P_{rWT} \left(\dfrac{v - v_i}{v_r - v_i} \right) & \text{for} \quad (v_i \leq v \leq v_r) \\ P_r & \text{for} \quad (v_r < v \leq v_o) \end{cases} \tag{3.3}$$

where C_n is the WT's operation cost. v_o, v_r, and v_i represent the cut-out, rated, and cut-in velocities of the WT. P_{rWT} is the nominal power of WT, while P_W is its output power. NWT refers to NO. WTs. Cost^{DG} can be obtained as

$$\text{Cost}^{\text{DG}} = \sum_{i=1}^{\text{NDG}} \alpha_i P_{DG,i,t}^2 + \gamma_i P_{DG,i,t} + \beta_i \tag{3.4}$$

where $P_{DG,i,t}$ denotes the outpower of the i-th DG at time t. α_i, γ_i, and β_i are cost parameters of the i-th DG. *NDG* is NO. DGs in MG. Finally, Cost^{FC} can be assigned as follows:

$$\text{Cost}^{FC} = \vartheta_n \sum_{i=1}^{NFC} \frac{P_{FC.i}}{\eta_{FC,i}} \tag{3.5}$$

where $P_{FC.i}$ is the FC's generated power of the i-th FC. η_{FC} and ϑ_n are the efficiency and cost of natural gas of the FC. Modeling the EV is associated with its battery, where its state of charging (E) can be described as follows:

$$E_i^t = E_i^{t-1} + \eta_{ch,i} \times P_{ch,i}^t \times \Delta t - \frac{\Delta t \times P_{dis,i}^t}{\eta_{dis,i}} \tag{3.6}$$

where P_{ch}^t and η_{ch} refer to the charring power and the efficiency of the charger, respectively. Likewise, P_{dis}^t and η_{dis} refer to the discharging power and efficiency of the discharging, respectively.

The system equality and inequality constraints can be represented as follows:

$$\sum_i^{NWT} P_{W,i}^t + \sum_i^{NFC} P_{FC,i}^t + \sum_i^{NDG} P_{DG,i}^t = P_{Load}^t + \sum_i^{NEV} P_{ch,i}^t - \sum_i^{NEV} P_{dis,i}^t \tag{3.7}$$

$$P_i^{\min} \leq P_{DGi} \leq P_i^{\max} \tag{3.8}$$

$$P_{FC}^{\min} \leq P_{FC} \leq P_{FC}^{\max} \tag{3.9}$$

$$P_{rWT}^{\min} \leq P_{rWT} \leq P_{rWT}^{\max} \tag{3.10}$$

$$E^{\min} \leq E \leq E^{\max} \tag{3.11}$$

$$P_{ch}^{\min} \leq P_{ch} \leq P_{ch}^{\max} \tag{3.12}$$

3.3 Dandelion optimizer

The DO simulates the motion of dandelion seeds across short distances. The motion of seeds is based on the wind speed and weather variations in which weather variations can control the seeds' flight pattern while the wind speed controls the traveling distance of these seeds. The motion of seeds includes rising phase motion, landing phase motion, and descending phase motion. The mathematical description of this optimizer is as follows.

3.3.1 Initialization

The initial locations of the dandelion seeds are randomly generated as follows:

$$X_i = Lo + r1 \times (Up - Lo) \tag{3.13}$$

where *Up* and *Lo* refer to the maximum and lower boundaries of variables, respectively. $r1$ is a random value between 0 and 1. Then, the corresponding objective function is evaluated for the generated locations, and the best solution is captured as follows:

$$f_{best} = \min(f(X_i)) \tag{3.14}$$

$$X_{best} = X(\text{find}(f_{best} == f(X_i))) \tag{3.15}$$

3.3.2   Rising stage

In this stage, the seeds will move and follow the change in weather and wind velocity. The seeds will move in long distances randomly, in which the lognormal distribution motion is used to describe this motion in a spiral as follows:

$$X(t+1) = X(t) + \alpha \times v_a \times v_b \times \ln Y \times (X_s - X(t)) \tag{3.16}$$

in which

$$X_s = Lo + r2 \times (Up - Lo) \tag{3.17}$$

$$v_a = r \times \cos\theta \tag{3.18}$$

$$v_b = r \times \sin\theta \tag{3.19}$$

$$r = \frac{1}{e^\theta} \tag{3.20}$$

$$\ln Y = \begin{cases} \dfrac{1}{y\sqrt{2\pi}}\exp\left[-\dfrac{1}{2\sigma^2}(\ln y)^2\right] & y \geq 0 \\ 0 & y < 0 \end{cases} \tag{3.21}$$

$$\alpha = r3 \times \left|\left(\frac{1}{T_{\max}^2}t^2 - \frac{2}{T_{\max}}t + 1\right)\right| \tag{3.22}$$

where $r2$ and α are random coefficients in which $0 \geq r2, \alpha \geq 1$. θ is a random angle in which $-\pi \geq \theta \geq \pi$. ln Y refers to a lognormal distribution. $T_{\max}$ and t denote the maximum and current iteration number. In this stage, the seeds move locally close to other neighborhoods as follows:

$$X(t+1) = X(t) \times (1 - \text{rand}()\times q) \tag{3.23}$$

where

$$q = \frac{1}{T_{\max}^2 - 2T_{\max} + 1}t^2 - \frac{2}{T_{\max}^2 - 2T_{\max} + 1}t + 1 + \frac{1}{T_{\max}^2 - 2T_{\max} + 1} \tag{3.24}$$

Finally, the raising stage can be represented as follows:

$$X(t+1) = \begin{cases} X(t) + \alpha \times v_a \times v_b \times \ln Y \times (X_s - X(t)) & \text{randn} < 1.5 \\ X(t) \times (1 - \text{rand} \times q) & \text{else} \end{cases} \tag{3.25}$$

where randn refers to a random coefficient that is generated randomly by a normal distribution.

3.3.3 Descending stage

In this stage, Brown motion is used to describe the seeds motion for the exploration process, and it can be expressed as follows:

$$X(t+1) = X(t) - \alpha \times r3 \times (X_{\text{mean}}(t) - \alpha \times r3 \times X(t)) \tag{3.26}$$

where $r2$ is a random coefficient in which $0 \geq r3 \geq 1$. X_{mean} is the mean position of the seeds, which are expressed as follows:

$$X_{\text{mean}}(t) = \frac{1}{\text{Npop}} \sum_{i=1}^{\text{Npop}} X_i \tag{3.27}$$

Npop denotes the number of populations.

3.3.4 Landing stage

In this motion, the locations of the seeds will be updated to be close to the elite solution as follows:

$$X(t+1) = X_{best}(t) + \text{Levy}(\tau) \times \alpha \times (X_{best}(t) - X(t) \times \varepsilon) \tag{3.28}$$

where $\text{Levy}(\tau)$ refers to the Levy flight function, and it can be formulated as follows:

$$\text{Levy}(\tau) = s \times \frac{r4 \times \sigma}{|r5|^{\frac{1}{\beta}}} \tag{3.29}$$

$$\sigma = \left(\frac{\Gamma(1+\beta) \times \sin\left(\frac{\pi\beta}{2}\right)}{\Gamma\left(\frac{1+\beta}{2}\right) \times \beta \times 2^{\left(\frac{\beta-1}{2}\right)}} \right) \tag{3.30}$$

where β and s represent the constant value that is equal to 1.5 and 0.01, respectively. $r4$ and $r5$ are random values in the range 0 and 1. ε denotes the adaptive coefficient, which can be expressed as follows:

$$\varepsilon = \frac{2t}{T_{\text{max}}} \tag{3.31}$$

3.4 Modified dandelion optimizer

The MDO is based on the integration of two improvement strategies, including the FDB and the local, for boosting the global exploration and exploitation searching abilities. The FDB is a searching mechanism that has been integrated with several optimization techniques to enhance its performance [30–36]. The FDB is mainly based on updating the locations of the population based on selecting the best score. The best score is calculated as a function of the distance between the elite and the current solution as well as the corresponding objective function values of the populations. The distance can be determined as follows:

$$D_i = \sqrt{\left(X_{i,1} - X_{best,1}\right)^2 + \left(X_{i,2} - X_{best,2}\right)^2 + \cdots + \left(X_{i,3} - X_{best,3}\right)^2} \quad (3.32)$$

The distance and the objective function matrix can be described as follows:

$$D_i = [D_1, D_2, D_3, \ldots, D_n] \quad (3.33)$$

$$Obj_i = [Obj_1, Obj, Obj_3, \ldots, Obj_n] \quad (3.34)$$

The score or the updated solution is based on FDB and can be calculated as follows:

$$Score_i = norm\ Obj_i + (1 - \rho) \times norm\ D_i \quad (3.35)$$

norm refers to the normalized value. ρ is a constant parameter that equals to 0.5. The second modification is the elite local motion, which has been proposed to enable populations to jump to be close to the best solution as follows:

$$X(t + 1) = X_{best}(t) + r1 \times \left(X_{t,i-1} - X_{t,i}\right) + \gamma \times \left(X_{best}(t) - X(t)\right) \quad (3.36)$$

in which

$$\gamma = 2 \times e^{r2((T_{max} - t + 1)/T_{max})} \times (\sin \times (2 \times r1 \times \pi)) \quad (3.37)$$

Figure 3.2 illustrates the key steps involved in applying the proposed MDO to solve the ED of a grid-connected hybrid MG, which incorporates both conventional and renewable energy resources to supply an electric load and an EV charging station. The MDO integrates FDB with a local elite movement strategy. The ED of the hybrid MG is addressed through the implementation of DSR and an intelligent EV smart charging strategy.

3.5   Results

The proposed MDO is developed for the ED solution of the MG, and the obtained results are compared with GWO [37], WOA [38], SCSO [39], and DO [22]. As mentioned before, the studied MG composites of one diesel generator, three FCs, and two WTs in which these units are used for an electric load demand and EVs charging station. The operation costs of these units are provided in Table 3.1. Additionally, the cost of purchased energy from the grid is displayed in Figure 3.3, while the load demand and the wind speed are displayed in Figures 3.4 and 3.5. The parameters of the competitive optimizers are listed in Table 3.2, and for a fair comparison, the selected maximum iterations and populations for all optimizers are 200 and 50, respectively. It should be highlighted here that the maximum power that is procured from the main grid is 800 kW. The maximum state of charging of the EV is 50 kWh, and the maximum charging power is 10 kW. In this study, the EVs will charge only within their allowable limits. The mean and the standard deviation of the initial state of charging are 50% and 15, respectively. In addition to that, the mean and standard deviation of the arrival time are 8 and 3 h, respectively. The initial states of charging of the EVs are displayed in Figure 3.6.

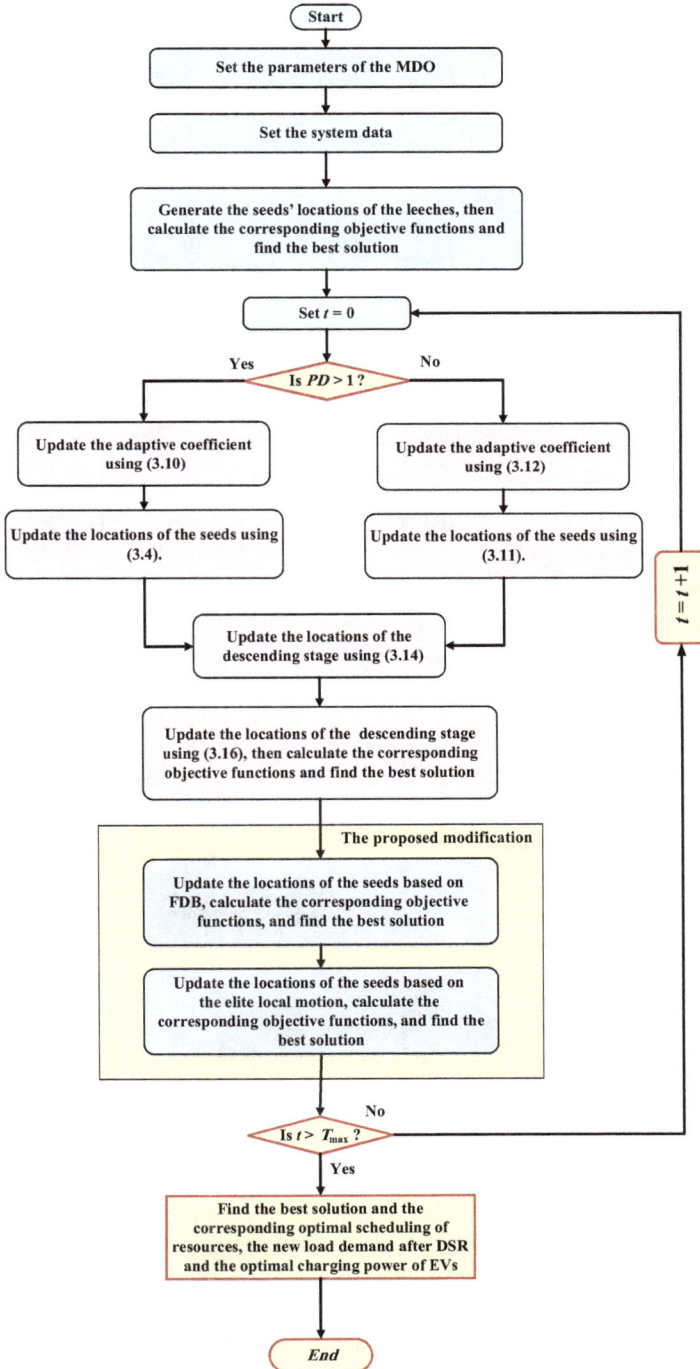

Figure 3.2 Flowchart for application of MOD in solving ED problem

Table 3.1   The cost data of the MG's components

Generation type	P^{max} (kW)	P^{min} (kW)	η_{FC} (%)	V_im (s)	V_om (s)	V_rm (s)	a_i ($/(kW)^{2}h)	γ_i ($/kWh)	β_i ($/h)
Diesel1	400	0	–	–	–	–	0.0074	0.2333	0.4333
FCell1	150	0	90	–			0	0.0500	0
FCell2	100	0	90	–	–	–	0	0.0500	0
FCell3	100	0	85	–	–	–	0	00.0700	0
Wind1	300	0	–	5	15	10	0	00.0220	0
Wind2	300	0	–	5	15	10	0	0.0320	0

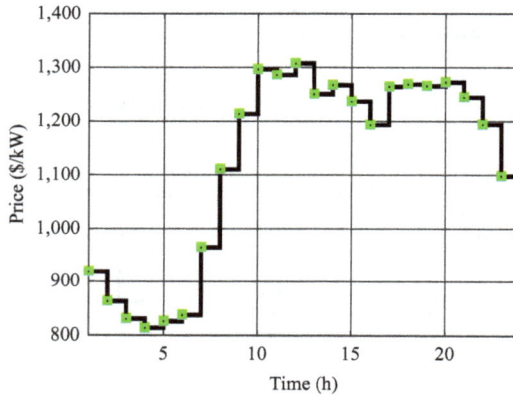

Figure 3.3   The cost of the purchased energy from the grid

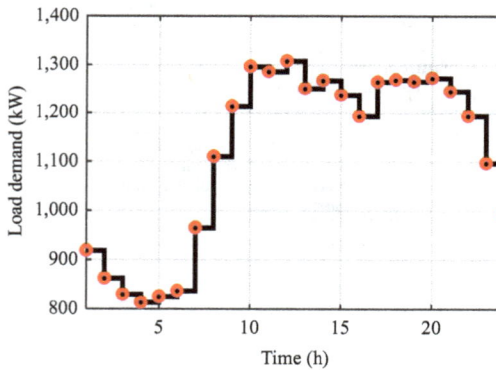

Figure 3.4   The expected load demand of the studied MG

3.5.1   Case 1: ED solution without a smart strategy

The ED of the MG was solved with the application of the smart strategy (DSR and smart charging strategy). The costs for ED that have been obtained by MDO, DO,

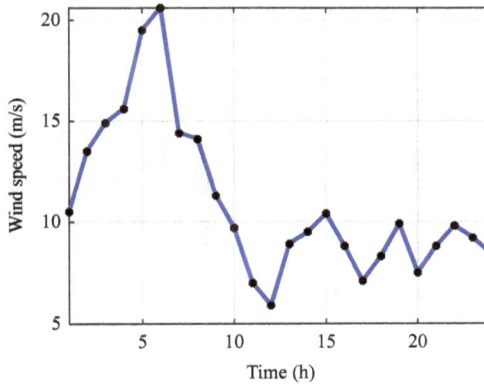

Figure 3.5   The expected daily wind speed

Table 3.2   The optimizers' parameters

Algorithm	Parameter	Value
MDO	α	[0,1]
	k	[0,1]
DO [40]	α	[0,1]
	k	[0,1]
GWO [37]	A	[2,0]
	C	2
WOA [38]	$a1$	[2,0]
	$a2$	[2,0]
	C	$2rand\,(0,1)$
	l	[−1,1]
	b	1
SCSO [39]	rg	[2,0]
	R	[−2rg, 2rg]

GWO, WOA, and SCSO are displayed in Table 3.2. The minimum obtained costs by MDO and DO are 16,590.23$ and 17,309.39$, respectively. According to Table 3.3, the best results can be obtained by application of the proposed MDO are the best for the means, the worst, and the best values. In this case, the EVs will charge at their maximum power capacity of 10 kW. The states of charging of EVs are depicted in Figure 3.6, while the number of EVs is illustrated in Figure 3.7. The optimal scheduling of the diesel, WTs, and FCs is displayed in Figure 3.8. Figure 3.9 shows the progress of charging of the EVs. It is clear that most EVs charge within 3–4 h only because all EVs are charged with a maximum charging power of 10 kW.

3.5.2   Case 2: the ED solution with a smart strategy

In this case, the MDO has been utilized for ED solution of MG with the application of the DSR program and smart charging strategy of the EVs. Table 3.4 lists the numerical

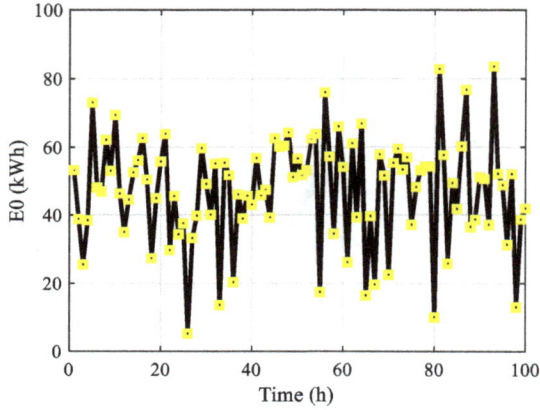

Figure 3.6   The initial state of charging of the EVs

Table 3.3   The costs for the competitive technique without a smart strategy

Optimizer	Mean	Best	Worst
SCSO	27,450.96	21,099.44	33,029.62
GWO	19,713.02	19,041.56	20,291.97
WOA	23,247.99	18,322.72	31,254.57
DO	17,458.81	17,309.39	17,603.32
MDO	16,899	16,590.23	17,141.86

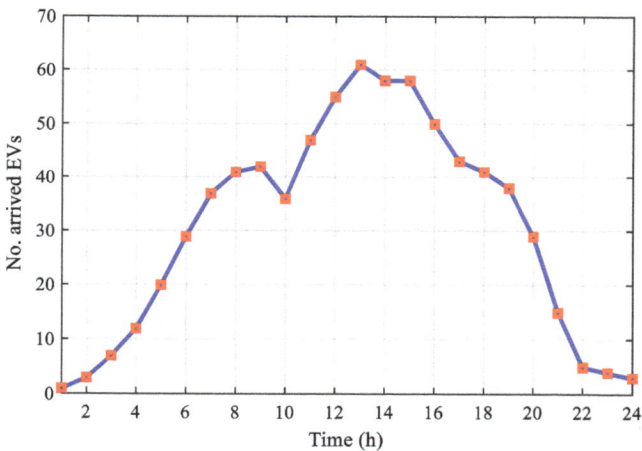

Figure 3.7   Percentage of the arrived EVs at a charging station

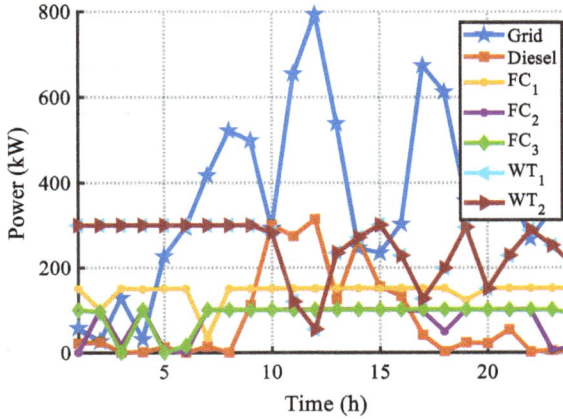

Figure 3.8   The optimal scheduling for the first case

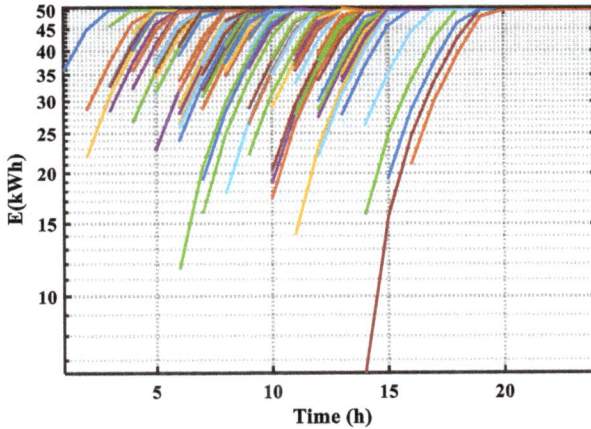

Figure 3.9   The state of charging EVs without a smarting strategy

results for costs that have been obtained by the optimization techniques studied. The minimum costs of MDO, DO, SCSO, GWO, and WOA are 10,884.43\$, 12,787.45\$, 18,254.95\$, 15,239.79\$, and 20,627.33\$, respectively. Thus, the lowest cost can be obtained by application of the proposed MDO. Furthermore, the cost was reduced from 16,590.23\$ (without smart strategy) to 10,884.43\$ with the application of the smart strategy. In other words, the cost was reduced by 34.3925% by the application of the smart strategy. The optimal scheduling of the MG's sources is listed in Figure 3.10. The states of charging of EVs at the case are displayed in Figure 3.11. It is obvious that the EVs take more time to charge allowable limits compared to the previous case. The load demands after and before the application of the smart strategy are shown in Figure 3.12. It is obvious that the load is shifted from 10:00 to 17:00 to other times because the price is high. Figure 3.13 displays the trend of the cost with the iterative

Table 3.4   The costs for the competitive technique with a smart strategy

Optimizer	Mean	Best	Worst
SCSO	21,207.3	18,254.95	29,504.18
GWO	17,043.46	15,239.79	17,810.04
WOA	23,097.09	20,627.33	25,015.62
DO	15,503.52	12,787.45	23,859.15
MDO	**12,128.29**	**10,884.43**	14,762.95

Note: The bold values refer to the best obtained results.

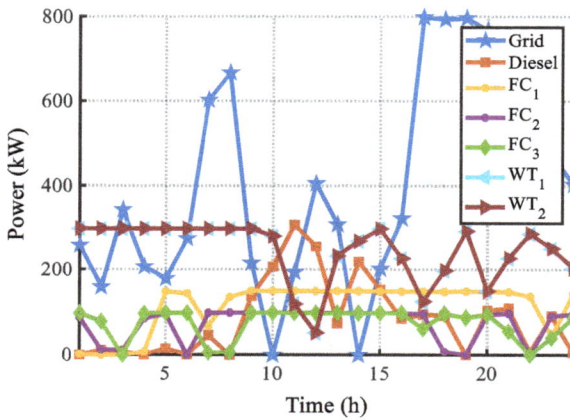

Figure 3.10   The optimal scheduling for the second case

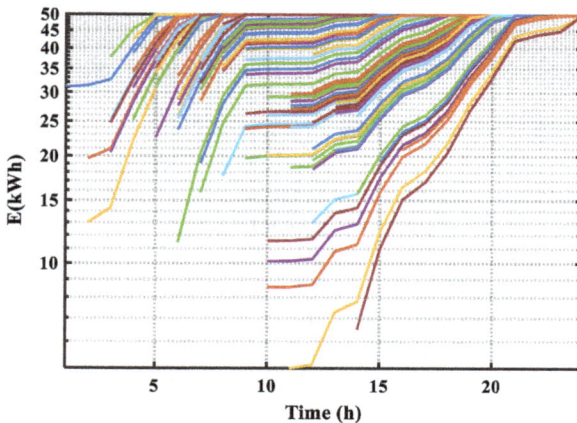

Figure 3.11   The state of charging of EVs with a smart strategy

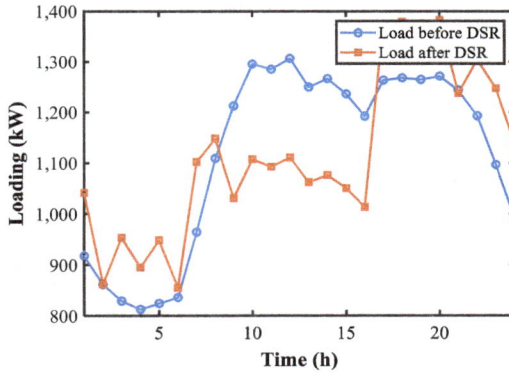

Figure 3.12   The system loads demand after and before the smart strategy

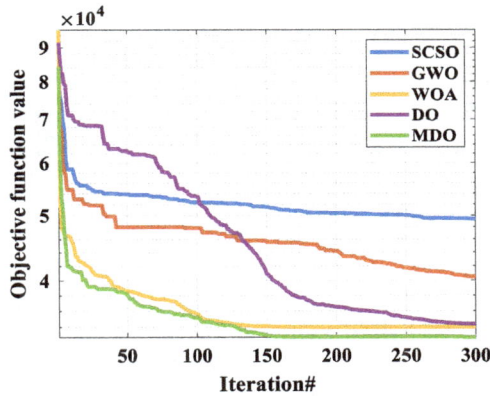

(a)

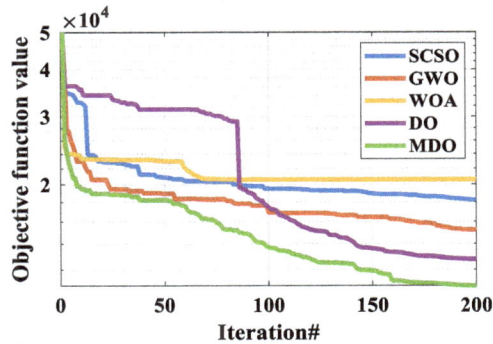

(b)

Figure 3.13   The trend of the objective function (a) without a smart strategy and (b) with a smart strategy

process against the other methods. It is clear that the MDO can converge compared to the optimal solution without fluctuations.

3.6   Conclusion

In this chapter, a novel modified DO optimizer was presented for solving the ED of an interconnected microgrid for reducing the operation cost. Two modifications were implemented to boost the performance of the suggested optimization methods including the FDB and the local elite solution. A smart strategy was proposed, which is based on the application of the DSR program and a controllable charging strategy for EVs. The results that have been obtained by the MDO with and without smart charging strategies have been compared with the obtained results by WOA, SCSO, GWO, and the traditional DO. As per the obtained results, the suggested MDO can solve the ED over the other competitive optimizer; furthermore, the application of the suggested smart strategy can reduce the operation cost by 34.3925%.

References

[1]   International Energy Agency (IEA), https://www.iea.org/reports/global-ev-outlook-2024/executive-summary [Accessed 20/2/2025].

[2]   O. J. Gbadeyan, J. Muthivhi, L. Z. Linganiso, and N. Deenadayalu, "Decoupling economic growth from carbon emissions: A transition toward low-carbon energy systems—A critical review," *Clean Technologies*, vol. 6, no. 3, pp. 1076–1113, 2024.

[3]   P. K. Saha, N. Chakraborty, A. Mondal, and S. Mondal, "Intelligent real-time utilization of hybrid energy resources for cost optimization in smart microgrids," *IEEE Systems Journal*, vol. 18, no. 1, pp. 186–197, 2024.

[4]   A. F. Güven, "Integrating electric vehicles into hybrid microgrids: A stochastic approach to future-ready renewable energy solutions and management," *Energy*, vol. 303, pp. 131968, 2024.

[5]   Y. Wu, Z. Wang, Y. Huangfu, A. Ravey, D. Chrenko, and F. Gao, "Hierarchical operation of electric vehicle charging station in smart grid integration applications—An overview," *International Journal of Electrical Power & Energy Systems*, vol. 139, pp. 108005, 2022.

[6]   A. P. Kaur, and M. Singh, "Time-of-use tariff rates estimation for optimal demand-side management using electric vehicles," *Energy*, vol. 273, pp. 127243, 2023.

[7]   F. Marzbani, and A. Abdelfatah, "Economic dispatch optimization strategies and problem formulation: A comprehensive review," *Energies*, vol. 17, no. 3, pp. 550, 2024.

[8]   Q. Hassan, S. Algburi, A. Z. Sameen, H. M. Salman, and M. Jaszczur, "A review of hybrid renewable energy systems: Solar and wind-powered solutions: Challenges, opportunities, and policy implications," *Results in Engineering*, pp. 101621, 2023.

[9] Z. Lin, C. Song, J. Zhao, and H. Yin, "Improved approximate dynamic programming for real-time economic dispatch of integrated microgrids," *Energy*, vol. 255, pp. 124513, 2022.

[10] M. Shadoul, R. Al Abri, H. Yousef, and A. Al Shereiqi, "Designing a dispatch engine for hybrid renewable power stations using a mixed-integer linear programming technique," *Energies*, vol. 17, no. 13, pp. 3281, 2024.

[11] G. Abbas, I. A. Khan, N. Ashraf, M. T. Raza, M. Rashad, and R. Muzzammel, "On employing a constrained nonlinear optimizer to constrained economic dispatch problems," *Sustainability*, vol. 15, no. 13, pp. 9924, 2023.

[12] R. A. Jabr, A. H. Coonick, and B. J. Cory, "A homogeneous linear programming algorithm for the security constrained economic dispatch problem," *IEEE Transactions on Power Systems*, vol. 15, no. 3, pp. 930–936, 2000.

[13] M. H. Hassan, S. Kamel, A. Selim, A. Shaheen, J. Yu, and R. El-Sehiemy, "Efficient economic operation based on load dispatch of power systems using a leader white shark optimization algorithm," *Neural Computing and Applications*, vol. 36, no. 18, pp. 10613–10635, 2024.

[14] W.-C. Yeh, M.-F. He, C.-L. Huang, *et al.*, "New genetic algorithm for economic dispatch of stand-alone three-modular microgrid in DongAo Island," *Applied Energy*, vol. 263, pp. 114508, 2020.

[15] M. Marhatang, and R. D. Muhammad, "Optimal economic dispatch using particle swarm optimization in Sulselrabar system," *International Journal of Artificial Intelligence*, vol. 11, no. 1, pp. 221–228, 2022.

[16] A. Kumar, M. Thakur, and G. Mittal, "Planning optimal power dispatch schedule using constrained ant colony optimization," *Applied Soft Computing*, vol. 115, pp. 108132, 2022.

[17] M. Basu, "Artificial bee colony optimization for multi-area economic dispatch," *International Journal of Electrical Power & Energy Systems*, vol. 49, pp. 181–187, 2013.

[18] S. Spea, "Optimizing economic dispatch problems in power systems using manta ray foraging algorithm: An oppositional-based approach," *Computers and Electrical Engineering*, vol. 117, pp. 109279, 2024.

[19] K. Nagarajan, A. Rajagopalan, M. Bajaj, R. Sitharthan, S. A. Dost Mohammadi, and V. Blazek, "Optimizing dynamic economic dispatch through an enhanced Cheetah-inspired algorithm for integrated renewable energy and demand-side management," *Scientific Reports*, vol. 14, no. 1, pp. 3091, 2024.

[20] Z. Xin-gang, L. Ji, M. Jin, and Z. Ying, "An improved quantum particle swarm optimization algorithm for environmental economic dispatch," *Expert Systems with Applications*, vol. 152, pp. 113370, 2020.

[21] X. Chen, and G. Tang, "Solving static and dynamic multi-area economic dispatch problems using an improved competitive swarm optimization algorithm," *Energy*, vol. 238, pp. 122035, 2022.

[22] S. Zhao, T. Zhang, S. Ma, and M. Chen, "Dandelion Optimizer: A nature-inspired metaheuristic algorithm for engineering applications," *Engineering Applications of Artificial Intelligence*, vol. 114, pp. 105075, 2022.

[23] A. E. Khalil, T. A. Boghdady, M. Alham, and D. K. Ibrahim, "A novel cascade-loop controller for load frequency control of isolated microgrid via dandelion optimizer," *Ain Shams Engineering Journal*, vol. 15, no. 3, pp. 102526, 2024.

[24] T. Ali, S. A. Malik, A. Daraz, S. Aslam, and T. Alkhalifah, "Dandelion optimizer-based combined automatic voltage regulation and load frequency control in a multi-area, multi-source interconnected power system with nonlinearities," *Energies*, vol. 15, no. 22, pp. 8499, 2022.

[25] A. Elhammoudy, M. Elyaqouti, D. B. Hmamou, *et al.*, "Dandelion Optimizer algorithm-based method for accurate photovoltaic model parameter identification," *Energy Conversion and Management: X*, vol. 19, pp. 100405, 2023.

[26] E. Halassa, L. Mazouz, A. Seghiour, A. Chouder, and S. Silvestre, "Revolutionizing photovoltaic systems: An innovative approach to maximum power point tracking using enhanced dandelion optimizer in partial shading conditions," *Energies*, vol. 16, no. 9, pp. 3617, 2023.

[27] M. Alharbi, M. Ragab, K. M. AboRas, *et al.*, "Innovative AVR-LFC design for a multi-area power system using hybrid fractional-order PI and PIDD2 controllers based on dandelion optimizer," *Mathematics*, vol. 11, no. 6, pp. 1387, 2023.

[28] S. Bergies, S.-F. Su, and M. Elsisi, "Model predictive paradigm with low computational burden based on dandelion optimizer for autonomous vehicle considering vision system uncertainty," *Mathematics*, vol. 10, no. 23, pp. 4539, 2022.

[29] M. Bastawy, M. Ebeed, A. Ali, M. F. Shaaban, B. Khan, and S. Kamel, "Optimal day-ahead scheduling in micro-grid with renewable based DGs and smart charging station of EVs using an enhanced manta-ray foraging optimisation," *IET Renewable Power Generation*, vol. 16, no. 11, pp. 2413–2428, 2022.

[30] H. T. Kahraman, S. Aras, and E. Gedikli, "Fitness-distance balance (FDB): A new selection method for meta-heuristic search algorithms," *Knowledge-Based Systems*, vol. 190, pp. 105169, 2020.

[31] S. Aras, E. Gedikli, and H. T. Kahraman, "A novel stochastic fractal search algorithm with fitness-distance balance for global numerical optimization," *Swarm and Evolutionary Computation*, vol. 61, pp. 100821, 2021.

[32] A. Adhikari, F. Jurado, S. Naetiladdanon, A. Sangswang, S. Kamel, and M. Ebeed, "Stochastic optimal power flow analysis of power system with renewable energy sources using adaptive lightning attachment procedure optimizer," *International Journal of Electrical Power & Energy Systems*, vol. 153, pp. 109314, 2023.

[33] M. Ebeed, A. Mostafa, M. M. Aly, F. Jurado, and S. Kamel, "Stochastic optimal power flow analysis of power systems with wind/PV/TCSC using a

developed Runge Kutta optimizer," *International Journal of Electrical Power & Energy Systems*, vol. 152, pp. 109250, 2023.

[34] A. T. Hachemi, F. Sadaoui, S. Arif, *et al.*, "Modified reptile search algorithm for optimal integration of renewable energy sources in distribution networks," *Energy Science & Engineering*, vol. 11, no. 12, pp. 4635–4665, 2023.

[35] M. Ebeed, S. Ali, A. M. Kassem, *et al.*, "Solving stochastic optimal reactive power dispatch using an Adaptive Beluga Whale optimization considering uncertainties of renewable energy resources and the load growth," *Ain Shams Engineering Journal*, vol. 15, no. 7, pp. 102762, 2024.

[36] M. Ebeed, M. A. Abdelmotaleb, N. H. Khan, *et al.*, "A modified artificial hummingbird algorithm for solving optimal power flow problem in power systems," *Energy Reports*, vol. 11, pp. 982–1005, 2024.

[37] S. Mirjalili, S. M. Mirjalili, and A. Lewis, "Grey wolf optimizer," *Advances in Engineering Software*, vol. 69, pp. 46–61, 2014.

[38] S. Mirjalili, and A. Lewis, "The whale optimization algorithm," *Advances in Engineering Software*, vol. 95, pp. 51–67, 2016.

[39] A. Seyyedabbasi, and F. Kiani, "Sand Cat swarm optimization: A nature-inspired algorithm to solve global optimization problems," *Engineering with Computers*, vol. 39, pp. 2627–2651, 2023.

[40] L. Wang, Q. Cao, Z. Zhang, S. Mirjalili, and W. Zhao, "Artificial rabbits optimization: A new bio-inspired meta-heuristic algorithm for solving engineering optimization problems," *Engineering Applications of Artificial Intelligence*, vol. 114, pp. 105082, 2022.

Chapter 4

Algorithms for techno-economic assessment of configuration and size of renewable energy systems

Hamdy M. Sultan[1,2], Ahmed S. Menesy[3] and Ahmed A. Zaki Diab[1,4]

Abstract

In this chapter, a novel isolated hybrid energy system is modeled and analyzed, with an emphasis on integrating biomass gasification technology to optimize system sizing and meet the electricity demand of a small, remote area with a peak load of 410 kW. The key components of the hybrid system include photovoltaic (PV) panels, wind turbines (WT), a biomass generators (BG), electrolyzer units, hydrogen storage tanks (HT), and a fuel cell (FC) system. To achieve the lowest possible energy cost while ensuring high reliability, measured through the loss of power supply probability (LPSP) and minimal power losses, the fata morgana algorithm (FATA), recently developed and inspired by geophysical principles, was employed. This optimization algorithm aims to meet all power demands while minimizing excess power (EXP) absorbed by the dump loads. A multi-objective function was constructed to address the optimization challenge, and the performance of the FATA algorithm was benchmarked against other techniques, including the RRT-based optimizer, artificial rabbits optimization, and hippopotamus optimization. The results demonstrated that FATA delivered the most effective system design, exhibiting superior convergence behavior compared to the other methods. Specifically, FATA achieved a fitness value of 0.064574, which corresponds to a system configuration consisting of 393 solar panels, 10 WTs, 3 biomass generators, a 399 kW electrolyzer, a hydrogen tank with a capacity of 103.59 kg, and a 145 kW FC. The system successfully provided electricity at an energy cost of 0.2620228 $/kWh, resulting in a net present cost of 6,326,140.79$, with an LPSP of 0.0183 and an EXP of 0.00608.

[1]Electrical Engineering Department, Faculty of Engineering, Minia University, Egypt
[2]Department of Mechatronics Engineering, Faculty of Engineering, Nahda University in Beni Suef, Egypt
[3]Electrical Engineering Department, King Fahd University of Petroleum and Minerals, Saudi Arabia
[4]Department of Mechatronics Engineering, Minia National University, Egypt

Keywords: Hybrid renewable energy system; Sizing; Optimization; Net present cost; Biomass generators; Fuel cell

4.1   Introduction

The adoption of clean and renewable sources of energy like wind [1], solar [2], geothermal [3], hydro [4], green hydrogen [5], and biofuels [6] has recently become an effective and environmentally friendly solution to address the global growth in energy demand and reduce the harmful effects caused by fossil fuels [7]. The increasing global demand for sustainable energy, coupled with the need to reduce carbon emissions, has heightened the focus on developing efficient energy systems. Isolated and rural areas, particularly in developing countries, face significant challenges in accessing reliable and affordable power due to geographical constraints and the high costs of extending conventional grid infrastructure [8]. In response to these challenges, hybrid microgrid systems have emerged as a promising solution, offering decentralized and renewable energy generation to meet local energy demands sustainably. Hybrid microgrids combine multiple energy sources, such as solar, wind, and biomass, with energy storage systems to ensure a reliable and continuous power supply [9,10]. Among these sources, biomass energy plays a crucial role, especially in rural and isolated communities where organic material is often readily available. Biomass can act as a stable and renewable source of power generation, particularly for base-load energy demands. Its integration into hybrid microgrids reduces reliance on fossil fuels, enhances energy security, and fosters local economic development by utilizing locally available resources. However, incorporating biomass into microgrids requires careful consideration of the system's technical and economic feasibility [11].

Developing an efficient hybrid microgrid system requires a comprehensive techno-economic analysis, which evaluates both the costs and performance of the system over its lifespan by considering technical configurations and economic impacts [12,13]. This analysis helps identify the most sustainable and cost-effective design choices for isolated communities by comparing different energy generation scenarios. Equally important is the optimal sizing of the system, where the capacities of generation units, storage, and biomass resources are carefully determined. Proper sizing ensures that the system avoids excessive costs or inefficiencies by balancing energy supply with demand, optimizing both investment and operational costs [14,15]. Optimization algorithms play a pivotal role in both the design and operation of hybrid microgrid systems, particularly in optimizing the sizing and scheduling of energy sources [14,16]. By applying these algorithms, decision-makers can find the best possible configuration that balances renewable energy supply, system reliability, and cost efficiency [4]. Optimization techniques help determine the ideal mix of energy sources, including biomass, solar, and storage, to meet energy demand in the most economically feasible way [17]. This chapter focuses on the techno-economic analysis and optimal sizing of hybrid microgrid systems, particularly considering the inclusion of biomass as a renewable energy source for isolated areas. By leveraging optimization algorithms, this study aims to enhance system performance and economic viability. Through detailed analysis, this chapter provides insights into the role of biomass in hybrid systems, the challenges of integrating it with other renewable sources, and the methods for optimizing the overall system design to efficiently meet the energy needs of remote communities.

The integration of renewable energy sources, including biomass, into hybrid microgrid systems presents a highly effective approach for sustainably addressing energy demands, particularly in isolated regions. Recent progress in optimization algorithms has greatly improved the performance, reliability, and economic viability of these systems by optimizing energy resource management and sizing configurations. This chapter will review recent studies that emphasize advanced optimization methods, renewable energy integration, and innovations in energy storage technologies, offering a detailed perspective on the latest research trends and developments in hybrid microgrids with a focus on systems that incorporate biomass for remote areas.

Islam *et al.* [18] explored the feasibility of a hybrid system incorporating solar photovoltaic (SPV), biomass generators (BG), diesel generators (DGs), and battery storage for rural electrification in the northern region of Bangladesh. Similarly, Shahzad *et al.* [19] and Kaur *et al.* [20] focused their research on providing optimal economic solutions for electrifying rural areas in Pakistan's Punjab province and India's northern Punjab state, respectively, using SPV/BG/battery configurations. Extending this work, Singh *et al.* [21], Li *et al.* [22], and Das *et al.* [23] further included wind turbines (WTs) into their system designs for remote villages in India, west China, and Bangladesh, thereby improving the hybrid system's reliability and renewable integration.

Meanwhile, Ahmad *et al.* [24] proposed a grid-connected SPV/BG/wind system without battery storage for Pakistan's Punjab province, highlighting different configurations for grid-tied systems. Additionally, the techno-economic viability of SPV/DG/battery systems has been studied by Lozano *et al.* [25], for the island of Cebu, Philippines, and Halabi *et al.* [26], for Malaysia. Several other studies further extended this system configuration by adding WTs for regions like Baluchistan, Sri Lanka, Iran, Colombia, and Ethiopia, as explored by Akram *et al.* [27], Kolhe *et al.* [28], Asrari *et al.* [29], Mamaghani *et al.* [30], and Gebrehiwot *et al.* [31]. These studies demonstrate the importance of wind energy in complementing SPV and biomass for hybrid systems in diverse geographical contexts. In addition to these developments, Ramesh and Saini [32] proposed an innovative hybrid power model for southern India by integrating small hydropower resources and exploring different dispatch strategies. Bhatt *et al.* [33] also considered hydro as a component in a hybrid SPV/BG/hydro/DG/battery system to address the energy needs of five energy-deprived villages in northern India. In a related study, Nag and Sarkar [34] assessed a hybrid SPV/hydro/BG/wind/battery system, emphasizing environmental sustainability. Similarly, Odou *et al.* [35] analyzed a hybrid SPV/hydro/DG/battery system for rural electrification in Benin. Despite these comprehensive studies, many still lack a focused approach to selecting energy sources based on local resource availability and performance for electricity generation, which is crucial for optimizing hybrid systems.

This chapter explores a novel approach to optimal sizing for an isolated hybrid energy system designed to meet the energy demands of a rural town, Abu-Monqar, located in Egypt's Western Desert. The proposed system integrates PV panels,

WTs, BG, electrolyzer units, hydrogen tanks (HTs), and fuel cells (FCs) as its main components. Previous studies focusing on the same case study area, such as the study by Diab *et al.* [36], employed a hybrid system comprising PV panels, WTs, diesel generators (DG), and batteries for optimal configuration. Additionally, Diab *et al.* [37] proposed a grid-connected system designed to minimize energy costs (EC) using a hybrid system with PV, WTs, and hydroelectric pumped storage. They utilized metaheuristic optimization algorithms like the whale optimization algorithm, water cycle algorithm, Salp swarm algorithm, and grey wolf optimizer to achieve their objectives. These studies highlight the effectiveness of using metaheuristic algorithms, inspired by natural phenomena, in solving complex optimization problems related to renewable energy systems. Despite their success, opportunities remain for enhancing existing methods or introducing new ones. This is aligned with the "no free lunch" theorem, which asserts that no single optimization method consistently outperforms others across all types of problems [38]. Therefore, in this chapter, four recent optimization techniques, fata morgana algorithm (FATA), RRT-based optimizer (RRTO), artificial rabbits optimization (ARO), and hippopotamus optimization (HO), are applied to solve the problem of optimal sizing for hybrid energy systems, incorporating PV, WT, BG, FC, electrolyzer, and HT units. The main contributions of this chapter could be summarized as follows:

1. A novel configuration has been designed, combining PV panels, WT, BG, FC, electrolyzer, and HT units. This system is aimed at improving the optimal size and reducing the cost of energy production for a small, isolated village in Egypt's Western Desert.
2. Developing a mathematical model to optimize the sizing of system components.
3. The newly introduced FATA, RRTO, ARO, and HO algorithms are applied to determine the optimal sizing of renewable energy sources in a stand-alone system. These methods aim to enhance the performance of the hybrid system by optimizing its design and configuration.

The structure of the chapter is as follows: Section 4.2 presents the design and configuration of the hybrid energy system. This section details the meteorological data for the case study area and explains the arrangement of system components such as PV, WT, BG, inverter, FC, electrolyzer, and HT units. Section 4.3 formulates the optimization problem, including the economic analysis, cost objectives, constraints, and operational strategies for the system. Section 4.4 introduces the FATA algorithm, explaining its mechanics and application in the context of the hybrid energy system. Section 4.5 discusses the results and key findings of the study, providing insights into the system's performance. Section 4.6 concludes the chapter, summarizing the major outcomes and potential areas for future research.

4.2 System modeling

Figure 4.1 illustrates the layout of the proposed stand-alone hybrid energy system. This system integrates PV panels, WTs, and BG as the primary sources of power

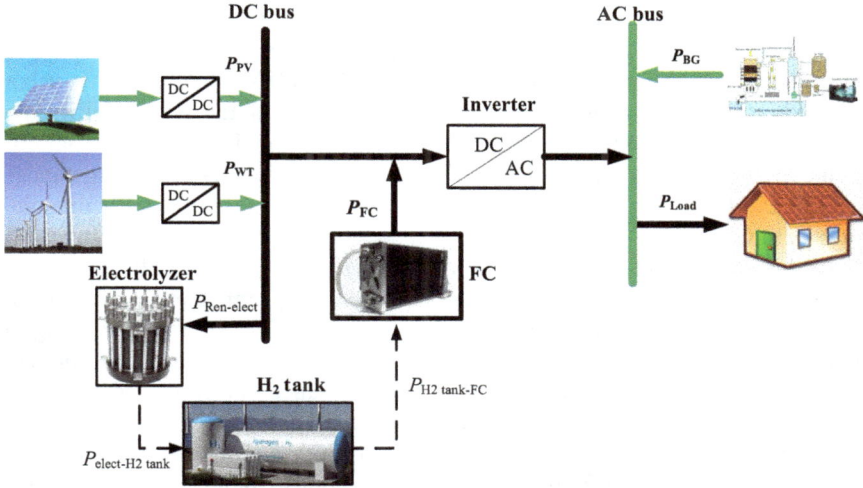

Figure 4.1   The schematic diagram of the proposed hybrid renewable energy system

generation. For energy storage, the system employs electrolyzers, hydrogen storage tanks, and FCs. These components work together to ensure a reliable energy supply, even during periods of low renewable generation. In off-grid hybrid systems like this one, managing surplus energy is crucial. To address situations where the HT is fully charged or there is an excess of generated power that is not immediately needed, dump loads are incorporated into the system. These dump loads act as a safety measure by dissipating the extra energy, preventing overcharging of storage components and ensuring the system remains stable and efficient. By combining multiple renewable energy sources with advanced energy storage solutions, this hybrid system aims to maximize the utilization of available resources while minimizing wastage. The inclusion of dump loads helps to optimize energy management and maintain the balance between power generation and consumption, particularly in remote or off-grid locations where reliable access to the energy grid is not possible.

4.2.1   PV system modeling

The performance of a solar module is influenced by various environmental and weather conditions at the installation site, causing fluctuations in the output power generated by the PV system [39]. The output power of the PV system (P_{pv}) can be determined using the following equations [40,41]:

$$T_{\text{cell}}(t) = R_{\text{int}}(t)\left(\frac{T_{NO} - 20}{0.8}\right) + T_{\text{amb}} \tag{4.1}$$

$$P_{pv}(t) = \left(N_{pv}p_{rated}^{PV}\eta_w\,\eta_{PV}\right)\left(\frac{R_{\text{int}}(t)}{R_{\text{nom}}}\right)(1 - \gamma_T(T_{\text{cell}}(t) - T_r)) \tag{4.2}$$

where $T_{\text{cell}}(t)$ represents the cell temperature at time (t), T_r represents the cell temperature under standard operating conditions, T_{NO} denotes the cell temperature under normal operating conditions, N_{pv} is the number of PV modules in the system, η_{PV} and η_w are the efficiency of the PV system and the wiring efficiency, respectively, p_{rated}^{PV} refers to the rated power of the PV module under normal conditions, R_{int} and R_{nom} are the solar radiation intensity at time (t) and under standard conditions, respectively, and γ_T is the temperature coefficient of the PV module's power output.

In this study, data on solar radiation and wind speed were collected for the proposed installation site, Abu-Monqar, based on long-term meteorological data provided by NASA. This dataset spans over 20 years of observation, ensuring the accuracy and reliability of the site's renewable energy potential. This approach enables precise analysis of the PV system's performance in response to real-world conditions, offering insights into the efficiency and optimization of hybrid energy systems for rural areas. By taking into account both environmental factors and system efficiencies, the proposed model offers a robust framework for predicting energy generation from solar resources at the selected location.

4.2.2   Wind farm modeling

A WT is composed of three main components: the tower, the blades, and the generator. These components work together to convert kinetic energy from the wind into electrical energy [42]. The amount of energy a WT can generate is influenced by the wind speed and the design of the blades [42,43]. Wind speed tends to increase with height above ground level, which means the wind speed at a particular height L_2 can be calculated using the following formula [43]:

$$\left(\frac{u_2}{u_1}\right) = \left(\frac{L_2}{L_1}\right)^{\beta_{wt}} \tag{4.3}$$

where u_1 and u_2 represent the wind speeds at the reference height L_1 and at the turbine hub height L_2, respectively, and β_{wt} is the WT's friction coefficient. According to the International Electrotechnical Commission (IEC) [44,45], the friction coefficient value varies depending on wind conditions. Under normal wind conditions, β_{wt} is typically 0.20, while in high wind scenarios, it is 0.11. The electrical power generated by a WT ($P_{wt}(t)$) is a function of the wind speed and is defined as follows [44,45]:

$$P_{wt}(t) = \begin{cases} 0, u(t) < u_{cut}^{in} \ or \ u(t) > u_{cut}^{off} \\ N_{wt}\, p_{rated}^{wt}\, \eta_{wt}\left(\dfrac{u^2(t) - u_{cut}^2}{u_{rated}^2 - u_{cut}^2}\right), u_{cut}^{in} < u(t) < u_{rated} \\ N_{wt}\, p_{rated}^{wt}\, \eta_{wt}, u_{rated} < u(t) < u_{cut}^{off} \end{cases} \tag{4.4}$$

where $u(t)$ is the wind speed at time (t), u_{cut}^{in} and u_{cut}^{off} are the cut-in and cut-off wind speeds, respectively, N_{wt} is the number of WTs in the system, η_{wt} represents the efficiency of the WT, and p_{rated}^{wt} is the rated power of the WT at the rated wind speed.

This set of equations highlights the dependency of WT power output on the wind speed, which varies throughout the day and across seasons. To optimize energy production, turbines are typically installed at a height where wind speeds are more consistent and stronger, ensuring a more reliable output of power. Furthermore, the friction coefficient plays an essential role in determining the rate of wind speed increase with height, directly influencing the turbine's performance and efficiency. Proper turbine siting and blade design are thus critical in maximizing energy capture and reducing losses.

4.2.3 Biomass system modeling

Biomass is a renewable and eco-friendly energy resource that contributes to reducing environmental pollution and mitigating global warming, especially when compared to fossil fuels, as its thermo-chemical conversion generates fewer harmful emissions [46,47]. In Egypt, large quantities of agricultural waste are often inefficiently managed, despite being a potential source for biomass energy [46]. These biomass materials can be converted into syngas, a synthetic gas, to generate electricity and heat through various thermodynamic conversion processes, such as gasification, direct combustion, and pyrolysis.

Among these techniques, biomass gasification holds significant promise, particularly for rural power generation, due to its lower cost relative to other conversion technologies [48]. The system proposed in this study utilizes a fixed-bed gasifier, operating on the gasification process, coupled with a gas turbine. Corn stover (CS), a common but underutilized agricultural residue in Egypt, serves as the biomass feedstock for the gasifier. During the gasification process, the biomass undergoes thermal decomposition at high temperatures exceeding 700°C. The breakdown of the biomass particles occurs in four stages, after which they react with a gasifying agent, typically air, to produce syngas. This syngas is subsequently used as fuel for gas turbines to generate electricity [4,49]. The efficiency of the syngas produced by the gasification process (η_g), the generator's output power ($P_{BG}(t)$), and the biomass fuel consumption rate per hour ($F_{Cons\_BG}(t)$) are determined using the following equations [48,49]:

$$\eta_g = \frac{LHV_{pg}m_{pg}}{LHV_B m_B} \tag{4.5}$$

where LHV_{pg} is the lower heating value of the product gas, m_{pg} is the mass flow rate of the product gas, LHV_B is the lower heating value of the biomass, and m_B is the mass flow rate of the biomass. The output power of the biomass generator at time t ($P_{BG}(t)$) is given by [47]

$$P_{BG}(t) = \frac{N_{BG}}{F_m} \left(\frac{\eta_g LHV_B \, Bio_{rated}(t)}{LHV_{pg}} - F_0 \, PG_{rated} \right) \tag{4.6}$$

where N_{BG} is the number of BG, Bio_{rated} is the biomass consumption rate per hour (kg/h), F_m represents marginal fuel consumption, F_0 is the no-load fuel consumption, and PG_{rated} is the rated power of the BG. The hourly biomass fuel consumption ($F_{Cons\_BG}(t)$) is calculated as [4,47]

$$F_{Cons\_BG}(t) = \frac{LHV_{pg}}{\eta_g LHV_B}(N_{BG}F_0 PG_{rated} + F_m P_{BG}(t)) \tag{4.7}$$

In this system, the conversion of biomass to syngas offers a sustainable and efficient means of producing electricity, particularly for remote or rural areas. Gasification is a superior option for handling agricultural waste, such as CS, which otherwise would remain unused. By optimizing the gasification process, including factors like biomass consumption and generator output, the system can achieve high efficiency while minimizing fuel use and environmental impact. The ability to produce electricity locally with a reduced carbon footprint highlights biomass gasification as a critical component of rural energy systems.

4.2.4   Electrolyzer modeling

An electrolyzer functions as a device that facilitates a chemical reaction by passing an electric current through a liquid medium. In the current investigation, a water electrolyzer is utilized to generate ultra-pure hydrogen in an environmentally friendly manner, with hydrogen gas being produced and stored at a pressure of 30 bar [50]. However, this hydrogen cannot be directly supplied to the FC since the reactant pressures within the FC can reach up to 1.2 bar [51]. To address this discrepancy, a hydrogen storage tank is implemented to connect directly to the electrolyzer [51,52]. The power transferred from the electrolyzer to the HT denoted as $P_{ELE/HT}$ can be represented by the equation:

$$P_{ELE/HT} = \eta_{ELE} * P_{ren/ELE} \tag{4.8}$$

In this equation, η_{ELE} signifies the efficiency of the electrolyzer, while $P_{ren/ELE}$ indicates the electrical energy supplied to the electrolyzer from renewable energy sources.

4.2.5   Hydrogen tank storage

Once the separation of hydrogen and oxygen is completed, hydrogen gas is transported to the HT for future utilization in the FC for energy generation. To maintain system balance, the upper pressure of the HT is calibrated to match the operational pressure of the water electrolyzer, while the lower pressure of the HT is aligned with that of the FC [53]. During its operation, the mass of hydrogen stored in the HT (HT_m) is regulated by specified minimum and maximum thresholds [54], as described by the following relationship:

$$HT_m^{min} \leq HT_m(t) \leq HT_m^{max} \tag{4.9}$$

The energy stored in the HT at a given time step $HT_P(t)$ and the hydrogen mass (HT_m) can be quantified using the equations:

$$HT_P(t) = HT_P(t-1)\left(P_{ELE/HT}(t) - \frac{P_{HT/FC}(t)}{\eta_{HT}}\right) * \Delta t \qquad (4.10)$$

$$HT_m(t) = HT_P(t)/HHV_H \qquad (4.11)$$

In these expressions, $P_{HT/FC}$ refers to the power transferred to the FC from the HT, η_{HT} represents the efficiency of the HT, Δt denotes the time interval during the simulation, and HHV_H indicates the higher heating value of the stored hydrogen gas (in kWh/m^3).

4.2.6 Fuel cell modeling

Fuel cells are electrochemical devices designed to convert chemical energy into electrical energy through continuous reactions involving hydrogen and oxygen gases. At the anode, hydrogen undergoes oxidation to produce protons, which migrate through the electrolyte to the cathode, while electrons flow externally to the anode. Upon collection, oxygen is reduced, leading to the formation of water. This study focuses on the polymer electrolyte membrane fuel cell (PEMFC), known for its compact design, durability, and efficient operation at relatively low temperatures [53]. The electrical power output generated by the PEMFC (FC$_P$) is dependent on its overall efficiency (η_{FC}), which is expressed as follows [51,52]:

$$FC_P = \eta_{FC}P_{HT/FC} \qquad (4.12)$$

By modeling these components—electrolyzer, hydrogen storage tank, and FC—this study aims to provide a comprehensive understanding of their interactions and efficiencies in a sustainable energy system. Technical and economic specifications of all components are listed in Table 4.1.

Table 4.1 Main parameters of the stand-alone hybrid system PV/biomass/FC

Component	Parameter	Value	Unit
PV system [36]	Model		PV-MLT260HC
	PV panel cost (PV$_{cost}$)	14,854	\$/m^2
	δ_T	0.0037	–
	PV$_\eta$	15	%
	T_r	25	°C
	Rated power	1	kW
	Length	1625	mm
	Width	1019	mm
	Thickness	46	mm
	lifetime of PV system (S$_{PV}$)	20	year
	PV replacement cost (C$_{rep}^{PV}$)	13885	\$

(Continues)

Table 4.1   (*Continued*)

Component	Parameter	Value	Unit
WT system [36]	Model	BWC Excel 10	
	Unit power	10	kW
	Cut-in speed	2.5	m/s
	Rated speed	13.8	m/s
	Cut-off speed	25	m/s
	Initial cost	1600	$/kW
	O&M cost	2	% of initial cost
	Replacement cost	1280	$/kW
	Life-time	20	year
	Efficiency	85	%
Inverter unit	η_{inv}	95	%
	Max. power	1	kW
	Inverter lifespan (S_{inv})	15	year
	Inverter capital cost	800	$/unit
	Replacement cost (C_{rep}^{inv})	750	$/kw
	O&M cost	8	$/unit-year
Gasification system [48,55]	LHV_B	14.8	MJ/kg
	LHV_{sy}	4.766	MJ/kg
	η_{sy}	80	%
	GP_{rat}	40	kW
	No-load fuel consumption (F_0)	0.0644	kg/h/50 kW
	Marginal fuel consumption (F_m)	0.2998	kg/h/50 kW
	Capital cost	23,700	$/kw
	Lifespan	15,000	h
	Replacement cost	15000	$/unit
	Yearly O&M cost	0.05	$/h
Electrolyzer unit [53]	Rated power	1	kW
	Lifespan	20	year
	Efficiency	75	%
	Capital cost	2000	$/unit
	Replacement cost	1500	$/unit
	Yearly O&M cost	25	$/unit
Hydrogen tank unit [53]	Power	1	kW
	Efficiency	95	%
	Capital cost	1300	$/unit
	O&M cost	15	$/unit
	Replacement cost	1200	$/unit
	Lifespan	20	year
FC system [53]	Power	1	kW
	Efficiency	50	%
	Capital cost	3000	$/unit
	O&M cost	175	$/unit
	Replacement cost	2500	$/unit
	Lifespan	5	year

4.3 Problem formulation

4.3.1 Cost analysis

The primary objective of this section is to optimize the proposed isolated hybrid energy system to ensure a reliable power supply while minimizing costs. This involves assessing key financial metrics, including total annual cost (C_{TA}), EC, and net present cost (NPC). The NPC represents the present value of the total capital investment and operating expenses over the lifespan of the project, as expressed by the formula:

$$\text{NPC} = C_{TA}/\text{CRF} \tag{4.13}$$

In this equation, C_{TA} refers to the total annualized cost, which aggregates the annualized costs of all subsystem components (C_A^y). The capital recovery factor (CRF) serves as a conversion tool to translate the initial investment cost into annual capital costs. The formulas for CRF and C_{TA} are provided below:

$$\text{CRF} = \frac{R(R+1)^S}{(R+1)^S - 1} \tag{4.14}$$

$$C_{TA} = \sum C_A^y = C_A^{PV} + C_A^{BG} + C_A^{ELE} + C_A^{HT} + C_A^{FC} + C_A^{inv} \tag{4.15}$$

Here, R denotes the interest rate, and S represents the expected lifespan of the hybrid system. The variable y includes various subsystem components, such as PV panels, BG, electrolyzer, HT, FC, and power inverter. Each component's annualized cost (C_A^y) can be calculated using the following formula:

$$C_A^y = C_{A\_Cap}^y + C_{OM}^y + C_{A\_Rep}^y + C_{A\_fuel} \tag{4.16}$$

In this context, $C_{A\_Cap}^y$ refers to the total annualized capital cost for each unit, C_{OM}^y accounts for operational and maintenance (O&M) costs, $C_{A\_Rep}^y$ pertains to replacement costs, and C_{A_fuel} denotes the annual fuel cost for the biomass unit. The eEC, measured in dollars per kilowatt-hour ($/kWh), reflects the expense associated with generating 1 kWh from the hybrid system and is calculated using the following expression:

$$\text{EC}\left(\frac{\$}{\text{kWh}}\right) = \frac{\text{NPC}(\$) * \text{CRF}}{\sum_1^{8760} \text{Load}_P(\text{kWh})} \tag{4.17}$$

4.3.2 Objective function

The goal of the optimization problem is to determine the ideal configuration of the system components to maximize the energy output of the hybrid renewable energy system while simultaneously minimizing the EC. Additionally, the objective includes ensuring high reliability of the power supply by reducing the loss of power supply probability (LPSP), represented by (4.19), and minimizing the excess energy P_{EXC} absorbed by the dummy load P_{Dum}. The P_{EXC} is defined in accordance with (4.21). The objective function is formulated as follows:

$$\text{Min}_X(F) = \text{Min}(\beta_1 \text{EC} + \beta_2 \text{LPSP} + \beta_3 \text{P}_{\text{EXC}}) \tag{4.18}$$

The LPSP is calculated using the following equation:

$$\text{LPSP} = \sum_1^{8760} \left(\frac{\text{Load}_P(t) - P_{\text{ren}}(t) - \text{FC}_P(t)}{\text{Load}_P(t)} \right) \tag{4.19}$$

The excess energy absorbed is expressed as follows:

$$P_{\text{EXC}} = \sum_1^{8760} (P_{\text{Dum}}/\text{Load}_P) \tag{4.20}$$

In these equations, β denotes the weighting factor for each objective, while X symbolizes the control variables associated with the optimization problem that need to be optimized through the applied algorithms:

$$X = \left[\text{PV}_N G_N \text{HT}_m P_{\underset{\text{HT}}{\text{ELE}}} \text{FC}_P \right] \tag{4.21}$$

4.3.3   Problem constraints

The optimization algorithm operates under specific constraints dictated by the upper and lower bounds of the decision variables:

$$1 \leq \begin{bmatrix} \text{PV}_N \\ WT_N \\ \text{BG}_N \\ P_{ELEC} \\ \text{FC}_P \\ \text{HT}_m \end{bmatrix} \leq \begin{bmatrix} \text{PV}_N^{\max} \\ WT_N^{\max} \\ \text{BG}_N^{\max} \\ P_{ELEC}^{\max} \\ \text{FC}_P^{\max} \\ \text{HT}_m^{\max} \end{bmatrix} \tag{4.22}$$

$$P_{\text{EXC}} \leq P_{\text{EXE}}^{\max} \tag{4.23}$$

$$\text{LSPS} \leq 0.06 \tag{4.24}$$

In these equations, PV_N, WT_N, and BG_N are restricted to integer values. The maximum allowable values for the components are defined as follows: $\text{PV}_N^{\max}$, $\text{WT}_N^{\max}$, and $\text{BG}_N^{\max}$ denote the upper limits for the number of PV panels, WTs, and gasifier generators, respectively. $P_{ELEC}^{\max}$ and $\text{FC}_P^{\max}$ refer to the maximum capacity in kilowatts of the electrolyzer and FC, respectively. Additionally, $\text{HT}_m^{\max}$ indicates the maximum storage capacity of the HT in kilograms.

4.3.4   Operation strategy

The proposed system functions as an isolated hybrid energy system designed to meet load demands. The operational strategy of this hybrid setup is illustrated in a flowchart (refer to Figure 4.2). The strategy can be categorized into three main scenarios:

1. Energy sufficiency: When the energy generated from renewable sources meets the load demand $P_{ren}(t) = \frac{\text{Load}_P(t)}{\eta_{inv}}$, the generated power is utilized to satisfy the required load, eliminating the need for input from the FC or electrolyzer.

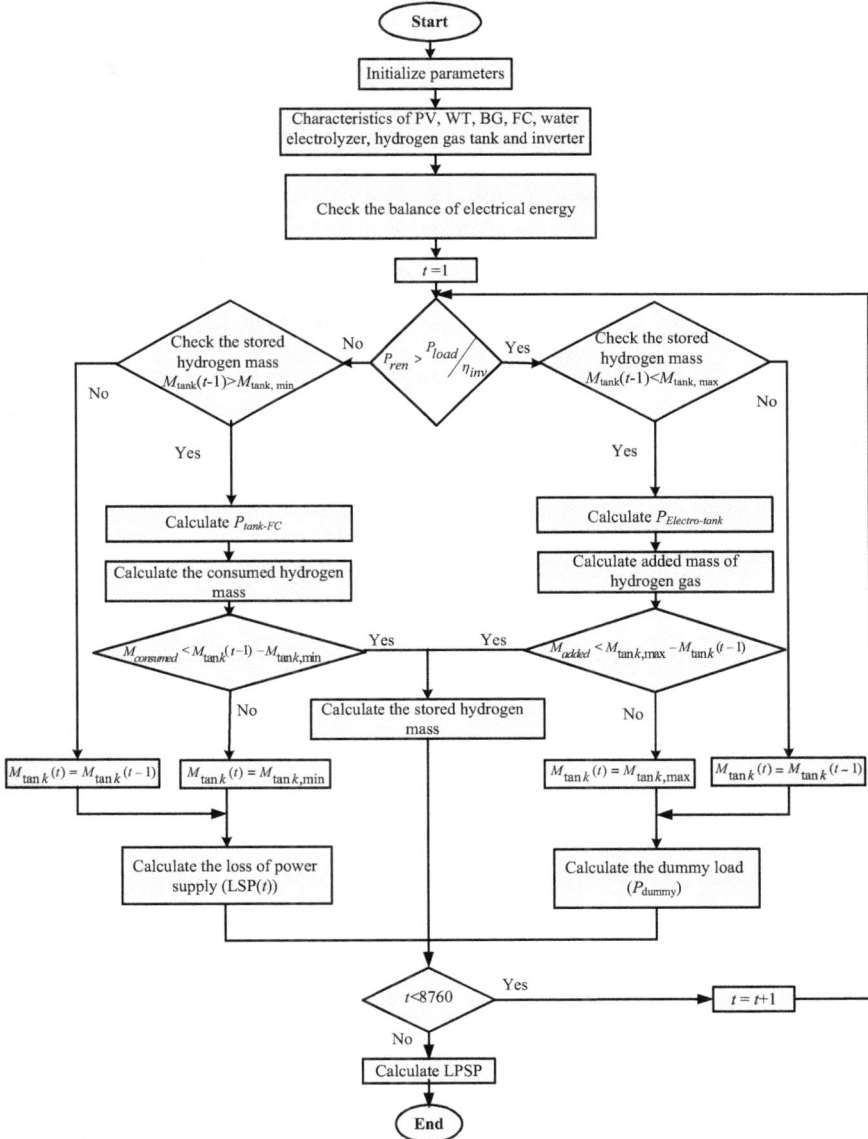

Figure 4.2   The flowchart of the proposed isolated energy system

2. Excess energy production: In cases where renewable energy output surpasses the load requirements $P_{ren}(t) > \frac{Load_P\ (t)}{\eta_{inv}}$, the surplus energy $P_{EXC}(t)$ is directed to the electrolyzer, producing hydrogen that is subsequently stored in the HT.
3. Insufficient energy generation: When renewable energy generation is inadequate to fulfill the load demand $P_{ren}(t) < \frac{Load_P\ (t)}{\eta_{iny}}$, the FC utilizes hydrogen stored in the HT to compensate for the shortfall in power generation. If the HT reaches its minimum capacity, this can lead to a loss of load.

4.4   FATA algorithm

The Fata Morgana, or mirage, is a frequently observed natural phenomenon caused by the propagation of light. It arises when light from an object is reflected through atmospheric layers of varying density, transitioning from a denser to a less dense medium. This chapter explores the mirages generated by light rays emanating from submerged hills and presents a design in Figure 4.3, which demonstrates how light rays from an ocean vessel can produce a mirage. The formation of a mirage requires both a medium with non-uniform density and the propagation of light through it. Initially, sunlight causes variations in atmospheric temperature, leading to the formation of this inhomogeneous medium. As light from the boat reflects into this medium, its refraction angle continuously changes during propagation, eventually resulting in total internal reflection and the appearance of the mirage. An observer (eye in Figure 4.3) can see the mirage by looking toward the sky in a red direction [56].

As illustrated in Figure 4.3, the complete mirage formation process shows that light emitted from a ship can only generate a mirage if a balance exists between filtering the mirage light and the refraction or reflection of light during propagation. Currently, as previously noted, an imbalance persists between global search and local search strategies in swarm intelligence algorithms for optimal value discovery. Inspired by the balanced interplay of mirage light filtering and light refraction/reflection in the mirage phenomenon, algorithms designed on this principle seek to harmonize global and local search strategies in optimization, thereby enhancing performance. By

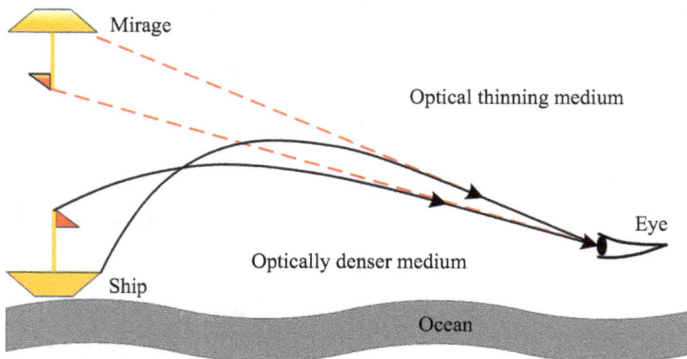

Figure 4.3   Mirage formation

contrast, algorithms such as the hippopotamus optimization (HO) mimic the global and local search strategies of swarm intelligence through a sequential application of soft and hard besiege strategies, as used in hawk hunting, but fail to sustain an effective balance [56]. Figure 4.3 examines how the FATA algorithm balances global and local search strategies using the mirage light propagation model. As light approaches the ship, it moves through a density gradient, shifting the refractive index and bending light at sharper angles until reaching a threshold where total internal reflection creates the mirage. FATA integrates these mirage-inspired mechanisms, introducing filtering to allow specific light paths to form the mirage and a propagation principle to adjust direction based on density, guiding light to an optimal reflection point.

In Figure 4.4, the multiple light sources generating a mirage in the Fata Morgana algorithm represent the population, while each individual light, denoted as light (*x*), targets optimization at the mirage (*x*_best). Initially, the population undergoes dynamic evaluation following the mirage light filtering principle, inspired by the definite integral principle. Rays emitted from the hull in the figure include those forming the mirage (*x*_best) and others directed elsewhere. In the second stage, the light population follows the light propagation strategy, utilizing initial and subsequent refraction, along with total internal reflection. These transformations through a medium of varying density enable information exchange, guiding the algorithm toward the optimal mirage target (solutions).

4.4.1   The mirage light filtering principle

This section introduces the population search strategy of the FATA algorithm, based on the principle of definite integrals. In Figure 4.4, during mirage formation, the hull emits two kinds of light rays: other light, which does not propagate and thus does not form a mirage, and the mirage light (*x*), which undergoes physical transformations that contribute to the mirage's creation.

In FATA, distinguishing between the two light populations is crucial for accurately identifying x_best. To facilitate this, FATA uses a quality evaluation strategy for the light population, grounded in the definite integral principle. In

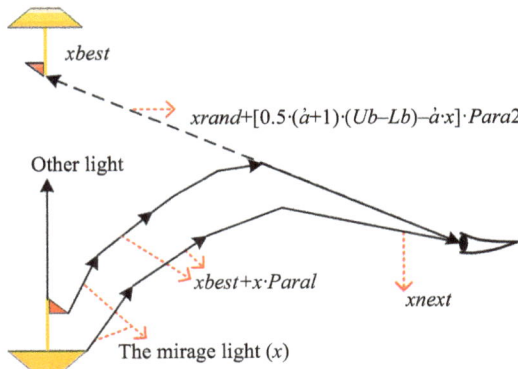

Figure 4.4   FATA optimization process in three dimensions

swarm intelligence algorithms, population quality is typically assessed by evaluating individual fitness and aggregating these across the population. As shown in Figure 4.5a, ranking individual fitness within a light population creates a cumulative curve. FATA then applies definite integration to this curve (Figure 4.5b) to efficiently calculate the fitness of both light types (other light and mirage light), using the integral as a fitness measure. The mirage light (x) selected through this approach is termed the filtered mirage light population.

The strategy initially classifies the population as either other light or mirage light based on overall population quality to apply distinct search methods (4.25). Here, population quality refers to the group's collective fitness. In this approach, the integrated area (S) of the population fitness function $(f(x))$ quantifies population quality, as shown in Figure 4.5a with the fitness function curve and in Figure 4.5b with the integrated area (S) beneath it. In SIA, fitness indicates individual quality, but assessing overall quality can be challenging with discrete, high-dimensional fitness values. Thus, individual fitness values are fitted to a function $(f(x))$, allowing the Fata Morgana algorithm to calculate the integrated area (S) of the population fitness curve using definite integration

$$x_i^{next} = \begin{cases} L_b + (U_b - L_b) \cdot rand, rand > P \\ x_{best} + x_i \cdot Para_1, rand \leq P \text{ and } rand < q \\ x_{rand} + [0.5 \cdot (a+1)(U_b - L_b) - ax_i] \cdot Para_2, rand \leq P \text{ and } rand \geq q \end{cases}$$
(4.25)

$$P = \frac{S - S_{worst}}{S_{best} - S_{worst}}$$
(4.26)

$$q = \frac{fit_i - fit_{worst}}{fit_{best} - fit_{worst}}$$
(4.27)

In this context, x represents the individual light, while x_i^{next} denotes the new individual. Equation (4.25) outlines the first-half refraction strategy, the second-half refraction strategy, and the total internal reflection strategy, respectively. In (4.26), P signifies the quality factor of the light population, where a smaller value of S indicates a higher quality of the population. v denotes the quality of the least fit

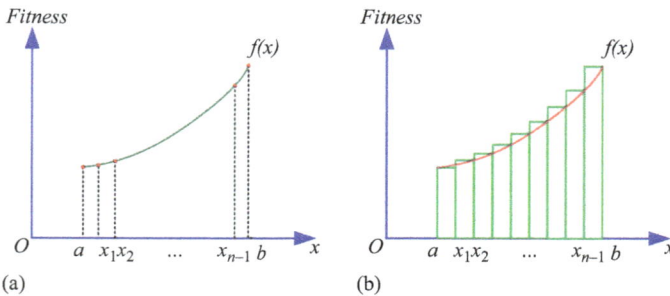

Figure 4.5   *The population fitness curve of FATA*

individual, while S_{best} indicates the quality of the most fit individual. The mirage light populations exhibit superior quality. In (4.27), q refers to the individual quality factor, fit_i represents the fitness of the current individual (x), fit_{worst} denotes the fitness of the least fit individual, and fit_{best} signifies the fitness of the most fit individual

$$y = f(x) = \sum_{j=0}^{n} c_j \varphi_j x \tag{4.28}$$

$$S = \int_a^b f(x)dx \approx \frac{b-a}{n} \cdot \left(\frac{y_0 + y_1}{2} + \frac{y_1 + y_2}{2} + \dots + \frac{y_{n-1} + y_n}{2} \right) \tag{4.29}$$

Equations (4.28) and (4.29) demonstrate the method for calculating the area under the population fitness curve $f(x)(f(x_1) < f(x_2) \dots < f(x_i) \dots < f(x_n))$ based on the principle of definite integration. This principle employs the concept of limits to compute the integrated area (S) of $f(x)$. Equation (4.29) represents the population quality fitting function $f(x)$, with points on the curve denoted as $x_i, y_i)$ and $i \in [1, n]$. c_j and φ_j are also included in this equation.

4.4.2 The light propagation principle

The light propagation principle in FATA follows the mirage light filtering principle, acting as the algorithm's individual search strategy to conduct local exploitation and locate local minima within the search space. As depicted in Figure 4.6, the

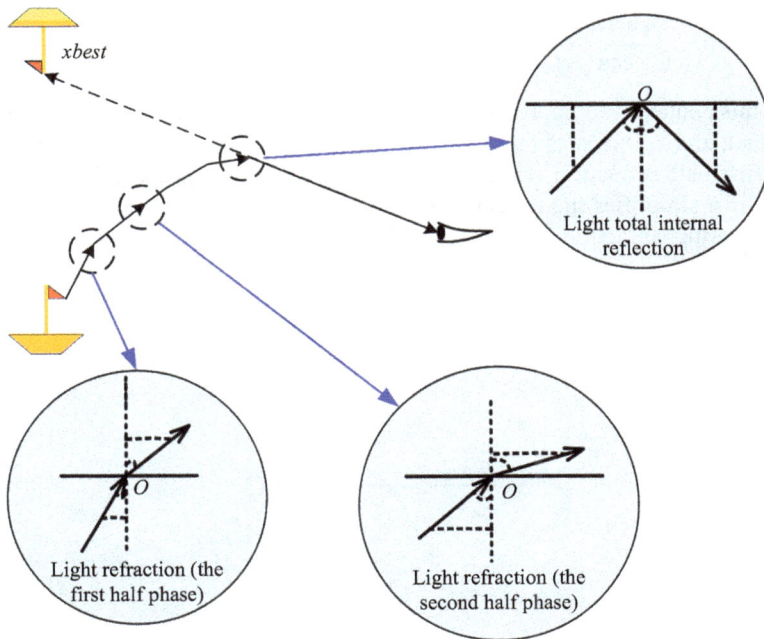

Figure 4.6 FATA is based on the mirage principle

FATA light population, represented by mirage light rays originating from the small boat, first undergoes filtering based on calculus principles to select rays contributing to the mirage. The filtered light population then experiences sequential refraction and reflection, shown in Figure 4.6 by changes in direction and size, guiding local search to identify local minima.

The Fata Morgana algorithm structures its individual search strategy on the light propagation principle, utilizing trigonometric functions. Based on the individual quality factor (outlined in (4.27)), the algorithm selects between the reflection strategy (first half phase), reflection strategy (second half phase), or the refraction strategy.

4.4.3   Light refraction (first half phase)

In Figure 4.7, light x enters a medium with variable density during the first phase of refraction, shifting from a denser to a less dense medium, which changes both its direction and size. The angle of incidence i_1 is smaller than the angle of refraction i_2. Figure 4.7 analyzes this refraction process for light individual x, where the level indicates the refractive surface. In (4.30), x^{next} represents the new individual formed through the first reflection phase, assuming $NO = C \cdot OM$, with C as a constant. Equations (4.30)–(4.32) provide the formulas for this strategy.

$$x^{next} = x_{best} + x_z \tag{4.30}$$

$$x_z = x \cdot Para_1 \tag{4.31}$$

$$Para_1 = \frac{\sin(i_1)}{C \cdot \cos(i_2)} = \tan(\theta) \tag{4.32}$$

In this context, x^{next} represents the new individual, x_{best} denotes the current best individual, and x_z indicates the refraction step of the strategy. The parameter $Para_1$ is the first-half refraction ratio, varying throughout light propagation. In (4.32), parameter θ simplifies the measurement of the incident angle i_1 and the reflection angle i_2 during refraction, with θ defined in the range [0,1].

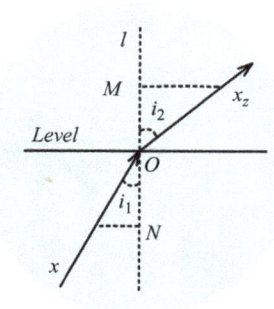

Figure 4.7   First refraction process of light

4.4.4   Light refraction (second half phase)

After the first refraction phase, the light undergoes a second refraction phase at random points. Figure 4.8 illustrates this second refraction process, where the incidence angle i_3 is less than the refraction angle i_3. As light propagates through a medium with varying density, the refractive index $Para_2$ changes continuously. In this phase, the light individual x_f generates a new individual x^{next} based on random individuals x^{next} within the search space. Equations (4.33)–(4.35) outline the formulas for this FATA strategy

$$x^{next} = x_{rand} + x_s \tag{4.33}$$

$$x_s = x_f \cdot Para_2 \tag{4.34}$$

$$Para_2 = \frac{\cos(i_5)}{C \cdot \sin(i_6)} = \frac{1}{\tan(\theta)} \tag{4.35}$$

where x_s represents the refraction stage in the second half refraction strategy, while x_{rand} is a random individual from the population. $Para_2$ denotes the second refraction ratio. To standardize $Para_1$ and $Para_2$, both parameters are scaled to the interval $[0, 1]$. The significant oscillation of $Para_2$ during the final phase of the FATA algorithm enhances its capability to avoid local optima.

4.4.5   Light total internal reflection

The total internal reflection phase marks the final stage of light propagation in the mirage formation process. As the refraction angle increases, the light undergoes total internal reflection within a medium of variable density, causing the FATA population to explore in the opposite direction. Figure 4.9 depicts this reflection process, where the incidence angle i_5 equals the reflection angle i_6. Here, x_0 represents the midpoint of the interval $[U_b, L_b]$, while E and F are the distances of the incident and refracted light to the horizontal plane, respectively. In this phase,

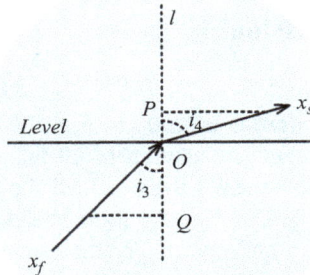

Figure 4.8   Second refraction process of light

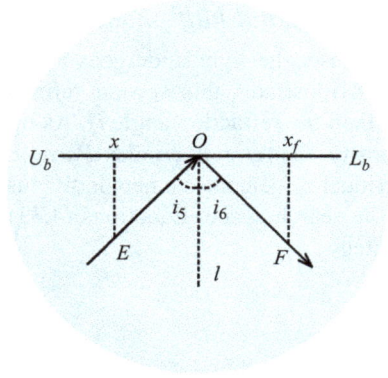

Figure 4.9   Total reflection process of light

the light individual x transforms into x^{next}, enabling reverse-direction target search. Equations (4.36)–(4.39) detail this FATA strategy

$$x^{next} = x_f = 0.5 \cdot (\alpha + 1)(U_b + L_b) - \alpha x \tag{4.36}$$

$$\alpha = \frac{F}{E} \tag{4.37}$$

$$x_0 - x_f = \frac{F \cdot (x - x_0)}{E} \tag{4.38}$$

$$x_0 = \frac{U_b - L_b}{2} + L_b = \frac{U_b + L_b}{2} \tag{4.39}$$

where x_f represents the individual reflected by strategy of the total internal reflection. α denotes the reflectance parameter of reflection strategy, which governs the transformation pattern of the light individual. When α is greater than 1, x^{next} crosses the boundary, with $\alpha \in [0, 1]$. U_b denotes the upper limit of the individual position, while L_b signifies the lower limit of the individual position.

4.5   Results and discussion

Abu-Monqar is a secluded settlement located in Egypt's Western Desert, approximately 100 km south of Farafra Oasis, 250 km from El-Dakhla Oasis, and 650 km southwest of Cairo. The geographical coordinates of this area are 26°30.3′N latitude and 27°39.8′E longitude. The system's average load for simulation purposes is 260 kW, with a peak demand of 420 kW. Figure 4.10 illustrates the hourly load distribution over both a single day and an entire year. For the case study, the monthly average values of global solar radiation and temperature were sourced from the NASA Surface Meteorology and Solar Energy database, relying on long-term averages collected over 20 years. Figures 4.11, 4.12, and 4.13 show the hourly average solar radiation on a horizontal plane, the hourly wind speed, and the corresponding temperature variations.

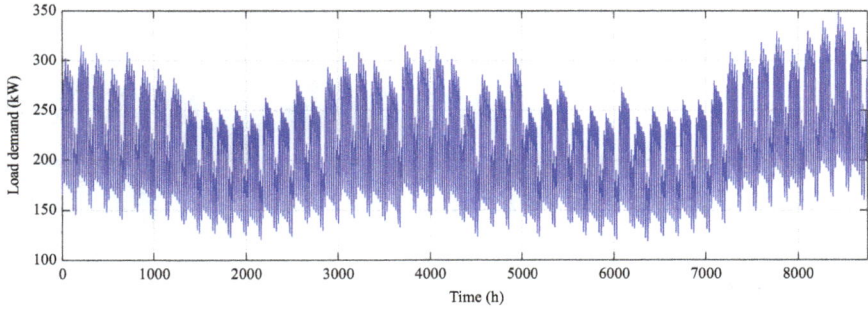

Figure 4.10   Annual load demand

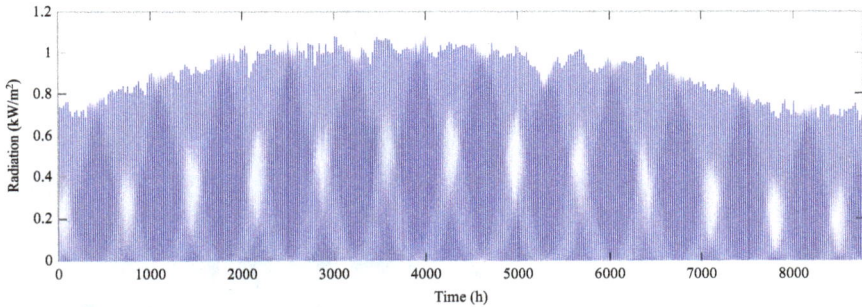

Figure 4.11   Hourly PV average radiation for one year

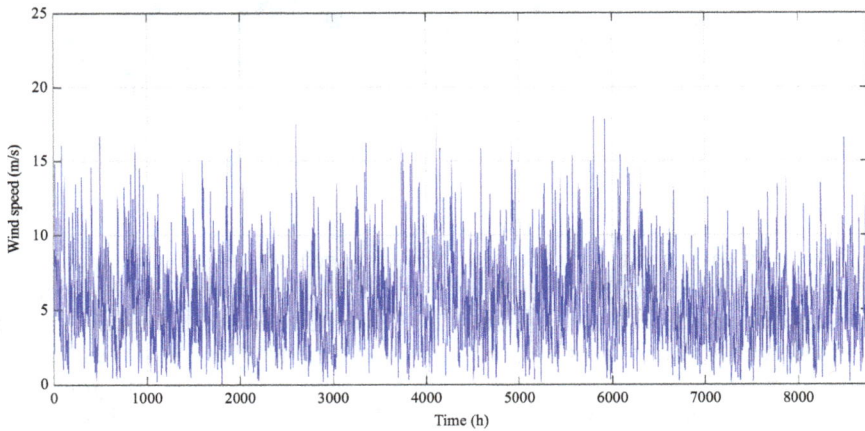

Figure 4.12   The hourly wind speed used during a year

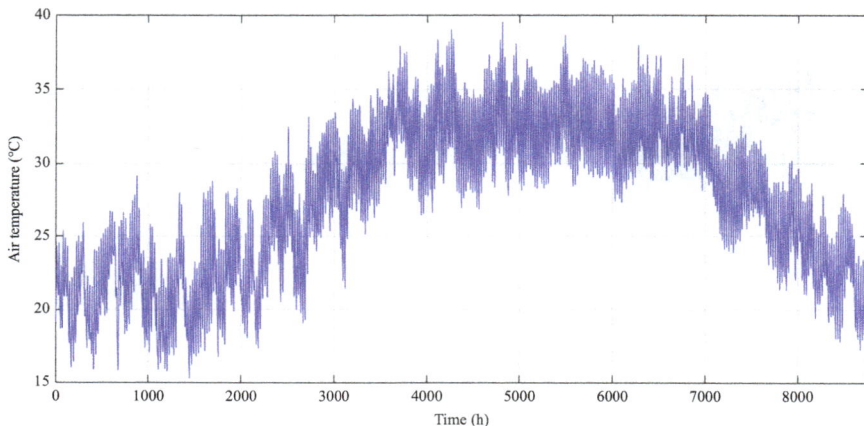

Figure 4.13   Hourly air temperature in the selected site for one year

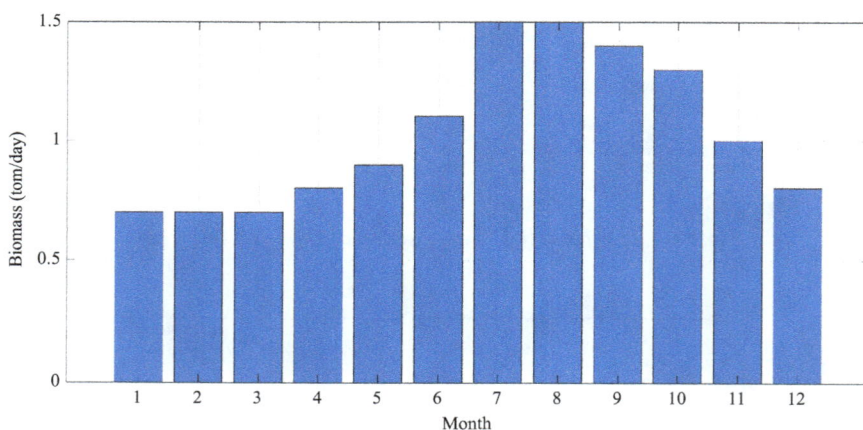

Figure 4.14   The rate of the feedstock used during a year

Biomass energy is a versatile renewable resource derived from organic materials, including plants or by-products from industries, agriculture, and households. This energy source can be converted into electricity or heat through various methods, such as thermolysis, anaerobic digestion, and gasification. Given Egypt's status as an agriculturally rich country, it has a significant supply of agricultural residues that can serve as biomass feedstock. Among these, rice straw is abundant in the northern regions, corn stalks in central Egypt, and sugarcane residues in the south. For this study, corn stalks were selected as the primary biomass feedstock due to their availability in proximity to the research area. Figure 4.14 shows the monthly consumption of the biomass feedstock used in the system.

In this study, a novel RRTO approach is proposed for determining the optimal sizing of a stand-alone renewable energy system, which includes components such as PV panels, WTs, biogas generators (BG), FCs, electrolyzers, and HTs. To validate the performance of the RRTO, its results are benchmarked against another advanced method, FATA, demonstrating RRTO's superior ability to achieve high reliability and cost-effectiveness.

All optimization algorithms were developed and executed in MATLAB. The specific parameters for each algorithm are as follows: population size = 20, maximum number of iterations (Max_it) = 200, and dimension size = 6.

The comparison between these methods highlights the robustness of FATA in optimizing system performance under various constraints. Figure 4.15 presents a comparison of the convergence curves for the different optimization algorithms. Based on these convergence characteristics, the FATA algorithm demonstrates superior performance, achieving a lower fitness value with faster convergence compared to other algorithms. This indicates that FATA not only reaches the optimal solution more efficiently but also minimizes the objective function more effectively. The enhanced convergence speed and reduced fitness value underscore FATA's advantage in handling the optimization process, making it a more reliable and cost-efficient approach for the system under consideration.

The results provided in Table 4.2 confirm the ability of all proposed algorithms in reaching the optimal solution to the optimization problem while keeping decision variables within their predefined limits. The power system reliability measured by the LPSP also kept below the upper threshold defined in the optimization constraints. Figure 4.16 shows the power distribution (P) from various energy sources (PV, WT, BG, ELECT, FC) over a 24-h period, compared against the power load demand. The x-axis represents time in hours, and the y-axis shows power in

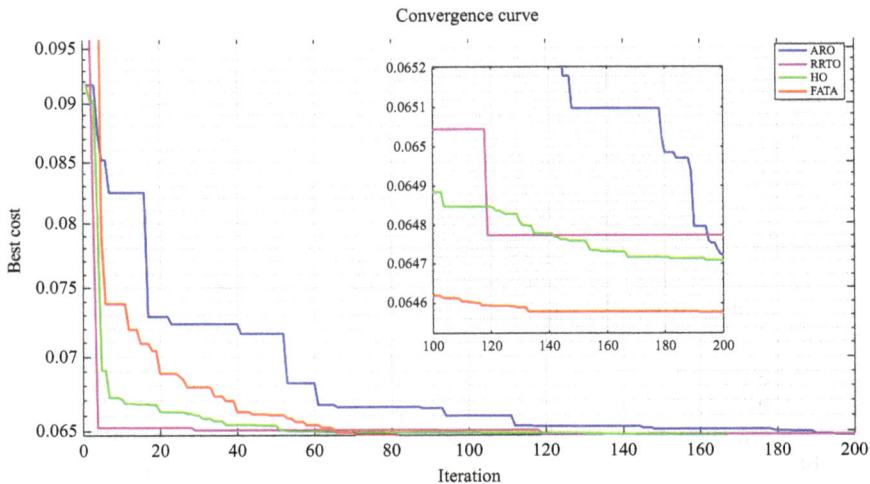

Figure 4.15   Convergence curves of RRTO and FATA

Table 4.2   The optimization results of ARO, HO, RRTO, and FATA techniques

	FATA	ARO	RRTO	HO
Best fitness function	0.064574224	0.064772810	0.06472153	0.064708105
Iteration number	132	199	119	193
PV (units)	392.17274	325.4115	446.781723	452.007064
WT (units)	9.516643	10.63911761	8.1989611	11.348062
Generators (units)	2.255360	2.2544213	2.242919	2.15800319
Electrolyzer rated (kW)	399.07293	388.613781	414.12093	400.217611
Hydrogen tank (kg)	103.59747	102.245759	106.92202	106.1291695
Fuel cell (kW)	144.79586	144.769773	147.86407591	145.2539635
EC ($/kWh)	0.259991	0.262022863	0.2664280400	0.2645589471
NPC ($)	6277103.494	6326140.788	6432497.021	6387370.635
LPSP	0.01825500	0.019601	0.017124171	0.0174031132
EXP	0.0060832	0.00506797	0.0058071351	0.0067722430

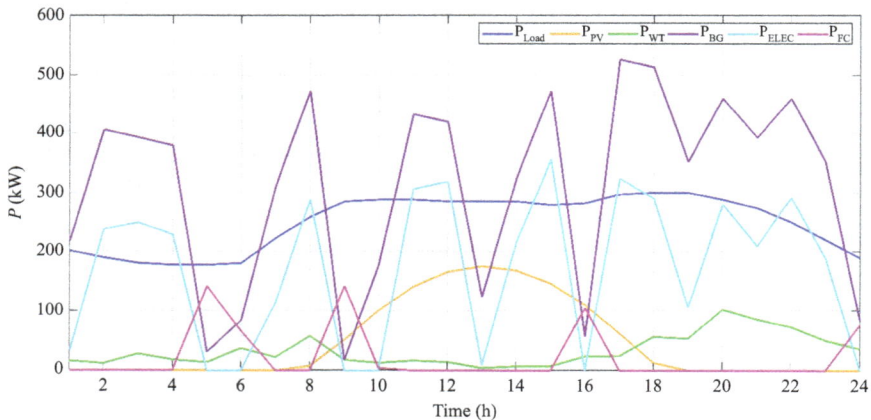

Figure 4.16   Power flow between all parts of the system in a certain day

kilowatts (kW). The different energy sources are indicated by distinct colored lines, with the following details: P_{Load} represents the system's power demand throughout the day, which gradually increases and fluctuates during peak hours; P_{PV} shows power generation from PV panels, peaking around midday when solar radiation is highest; P_{WT} denotes WT power generation, which appears to fluctuate significantly throughout the day, showing no distinct pattern; P_{BG} represents biogas generator output, providing a relatively steady contribution over time; P_{ELEC} shows the power drawn by the electrolyzer system, which also fluctuates depending on the available renewable resources; P_{FC} represents FC power generation, peaking during higher load demand periods, particularly in the morning and evening hours. The graph highlights the dynamic interaction between the different energy sources and the load demand, with each source contributing at varying levels throughout the day to meet energy needs.

Figure 4.17 displays a long-term power analysis over several thousand hours (likely representing nearly a full year of data), comparing load demand, power deficit, and power surplus: Load (blue line): This line shows the power demand across time, fluctuating regularly with peaks and troughs. The load appears to remain within a band of around 150–300 kW, reflecting a relatively consistent demand pattern. The power deficit (P_{def}) indicates periods when the energy generated by the system falls short of meeting the load demand. The power deficit appears sporadically, with some intervals experiencing significant shortfalls (up to around 100 kW). The power surplus (P_{sur}) shows times when the system generates more energy than required. The surplus is generally lower in magnitude compared to the deficit, with smaller peaks, indicating that excess energy is not consistently generated in large quantities. Overall, this graph reflects the system's balance between energy generation and consumption over time, with visible periods of both energy shortfalls and occasional surpluses, highlighting the system's variability in matching load demand with renewable energy production.

Figure 4.18 illustrates the variation in hydrogen mass over time, likely spanning nearly a full year based on the x-axis, which is labeled in hours. The mass fluctuates significantly, ranging from around 20 kg to slightly over 100 kg. The graph indicates dynamic changes in hydrogen production or storage, with peaks and troughs occurring frequently over time. The general trend suggests a periodic increase in hydrogen mass, followed by periods of depletion. This pattern could be influenced by the system's operation, where hydrogen is produced during times of surplus renewable energy (e.g., when solar or wind power exceeds demand) and consumed or stored when needed. Peaks appear more frequently around the middle of the timeline and seem to decrease toward the end, potentially reflecting seasonal variations or operational changes in energy production and consumption.

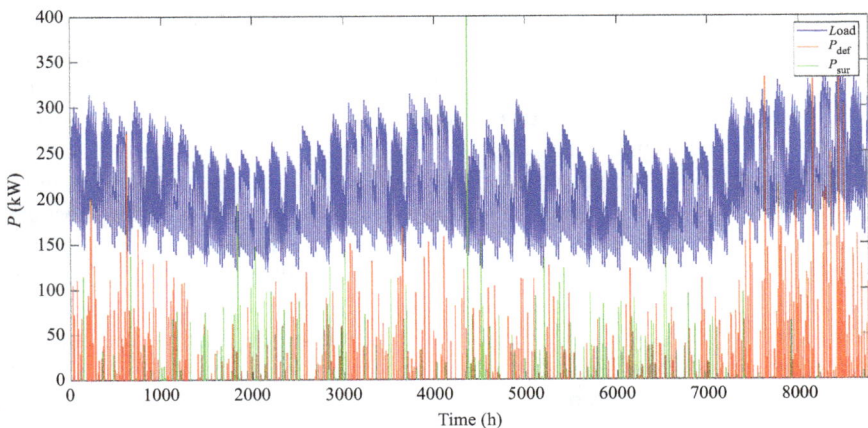

Figure 4.17   Loss of power and power dissipated in dummy load over the year

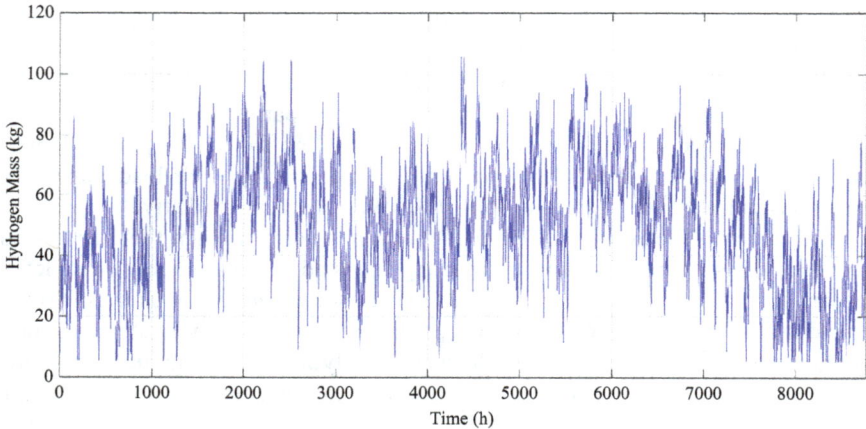

Figure 4.18   Variation of the mass of hydrogen in the tank over the year

4.6   Conclusion

The primary goal of this study is to assist policymakers, decision-makers, and the private sector in identifying the benefits, opportunities, and challenges surrounding the future of solar and biomass energy in Egypt. This includes analyzing trends in production costs from various technologies and exploring potential profit opportunities to attract private-sector investments, thereby encouraging the expansion of solar and biomass energy as viable alternatives to conventional energy sources.

In this work, four optimization techniques, namely, FATA, RRTO, HO, and ARO—are employed and their results are compared to propose an optimal design for a small, stand-alone microgrid capable of meeting the energy needs of a remote area in Egypt's Western Desert, which has a peak load of 410 kW. The hybrid energy system is based on three renewable resources: PV panels, WT, and a biomass gasification system. These are integrated with an FC and an electrolyzer for hydrogen production. The objective is to minimize the energy production cost while ensuring a high level of power supply reliability. Simulation results indicate that the proposed optimization methods are highly effective in determining the optimal capacities of the power generation and energy storage components in this off-grid hybrid system. Among the techniques, FATA proved to be the most efficient, delivering the best overall solution with a minimum energy cost of $0.2620228 per kWh, an NPC of $6,326,140.79, an LPSP of 0.0183, and an EXP ratio of 0.00608. Future studies could expand on this research by exploring different configurations and including additional renewable energy resources like geothermal or tidal power to further diversify the energy mix. Moreover, considering the uncertainties in renewable energy resources and fluctuating fuel prices, stochastic or probabilistic approaches could be incorporated into the optimization framework. Another avenue for future work would be the development of advanced control strategies for real-time operation of the system, as well as analyzing the economic and environmental impacts of scaling up the system to serve larger areas or regions. Finally, a deeper exploration of the integration of energy storage technologies beyond hydrogen, such as advanced battery systems, could be considered to enhance the overall performance and reliability of hybrid energy systems.

References

[1] S. Rehman, K. M. Kotb, M. E. Zayed, *et al.*, "Techno-economic evaluation and improved sizing optimization of green hydrogen production and storage under higher wind penetration in Aqaba Gulf," *Journal of Energy Storage*, vol. 99, p. 113368, 2024.

[2] S. Rehman, M. E. Zayed, K. Irshad, *et al.*, "Design, commissioning and operation of a large-scale solar linear Fresnel system integrated with evacuated compound receiver: Field testing, thermodynamic analysis, and enhanced machine learning-based optimization," *Solar Energy*, vol. 278, p. 112785, 2024.

[3] H. Yin, M. E. Zayed, A. S. Menesy, J. Zhao, K. Irshad, and S. Rehman, "Present status and sustainable utilization of hydrothermal geothermal resources in Tianjin, China: a critical review," *AIMS Geosciences*, vol. 9, no. 4, pp. 734–753, 2023.

[4] A. S. Menesy, H. M. Sultan, I. O. Habiballah, H. Masrur, K. R. Khan, and M. Khalid, "Optimal configuration of a hybrid photovoltaic/wind turbine/biomass/hydro-pumped storage-based energy system using a heap-based optimization algorithm," *Energies*, vol. 16, no. 9, p. 3648, 2023.

[5] A. S. Menesy, H. M. Sultan, M. E. Zayed, *et al.*, "A modified slime mold algorithm for parameter identification of hydrogen-powered proton exchange membrane fuel cells," *International Journal of Hydrogen Energy*, vol. 86, pp. 853–874, 2024.

[6] L. Jing, J. Zhao, H. Wang, *et al.*, "Numerical analysis of the effect of swirl angle and fuel equivalence ratio on the methanol combustion characteristics in a swirl burner," *Process Safety and Environmental Protection*, vol. 158, pp. 320–330, 2022.

[7] A. E. Geweda, M. E. Zayed, M. Y. Khan, and A. B. S. Alquaity, "Mitigating CO_2 emissions: A review on emerging technologies/strategies for CO_2 capture," *Journal of the Energy Institute*, vol. 118, p. 101911, 2025.

[8] B. Brand and T. Fink, "Renewable energy expansion in the MENA region: a review of concepts and indicators for a transition towards sustainable energy supply," *4th Asia-Africa Sustainable Energy Forum*, 2014.

[9] W. Cai, X. Li, A. Maleki, *et al.*, "Optimal sizing and location based on economic parameters for an off-grid application of a hybrid system with photovoltaic, battery and diesel technology," *Energy*, vol. 201, p. 117480, 2020.

[10]   W. Zhang, A. Maleki, M. A. Rosen, and J. Liu, "Sizing a stand-alone solar-wind-hydrogen energy system using weather forecasting and a hybrid search optimization algorithm," *Energy Conversion and Management*, vol. 180, pp. 609–621, 2019.

[11]   P. Kumar, N. Pal, and H. Sharma, "Optimization and techno-economic analysis of a solar photo-voltaic/biomass/diesel/battery hybrid off-grid power generation system for rural remote electrification in eastern India," *Energy*, vol. 247, p. 123560, 2022.

[12]   T. F. Agajie, A. Ali, A. F. Lele, *et al.*, "A comprehensive review on techno-economic analysis and optimal sizing of hybrid renewable energy sources with energy storage systems," *Energies*, vol. 16, no. 2, p. 642, 2023.

[13]   H. M. Sultan, A. S. Menesy, S. Kamel, A. S. Alghamdi, and M. Zohdy, "Optimal sizing of isolated hybrid PV/WT/FC system using manta ray foraging optimization algorithm," *International Transaction Journal of Engineering, Management, & Applied Sciences & Technologies*, vol. 11, no. 16, pp. 1–12, 2020.

[14]   M. Heidari, M. Heidari, A. Soleimani, *et al.*, "Techno-economic optimization and Strategic assessment of sustainable energy solutions for Powering remote communities," *Results in Engineering*, vol. 23, p. 102521, 2024.

[15]   H. M. Sultan, A. S. Menesy, S. Kamel, A. S. Alghamdi, and C. Rahmann, "Optimal design of a grid-connected hybrid photovoltaic/wind/fuel cell system," in *2020 IEEE Electric Power and Energy Conference (EPEC)*, 2020, pp. 1–6: IEEE.

[16]   H. Abd El-Sattar, H. M. Sultan, S. Kamel, A. S. Menesy, and C. Rahmann, "Optimal design of hybrid stand-alone microgrids using tunicate swarm algorithm," in *2021 IEEE International Conference on Automation/XXIV Congress of the Chilean Association of Automatic Control (ICA-ACCA)*, 2021, pp. 1–6: IEEE.

[17]   A. S. Menesy, S. Almomin, H. M. Sultan, *et al.*, "Techno-economic optimization framework of renewable hybrid photovoltaic/wind turbine/fuel cell energy system using artificial rabbits algorithm," *IET Renewable Power Generation*, vol. 18, no. 15, pp. 2907–2924, 2024.

[18]   M. S. Islam, R. Akhter, and M. A. Rahman, "A thorough investigation on hybrid application of biomass gasifier and PV resources to meet energy needs for a northern rural off-grid region of Bangladesh: A potential solution to replicate in rural off-grid areas or not?" *Energy*, vol. 145, pp. 338–355, 2018.

[19]   M. K. Shahzad, A. Zahid, T. ur Rashid, M. A. Rehan, M. Ali, and M. Ahmad, "Techno-economic feasibility analysis of a solar-biomass off grid system for the electrification of remote rural areas in Pakistan using HOMER software," *Renewable Energy*, vol. 106, pp. 264–273, 2017.

[20]   M. Kaur, S. Dhundhara, Y. P. Verma, and S. Chauhan, "Techno-economic analysis of photovoltaic-biomass-based microgrid system for reliable rural electrification," *International Transactions on Electrical Energy Systems*, vol. 30, no. 5, p. e12347, 2020.

[21] S. Singh, M. Singh, and S. C. Kaushik, "Feasibility study of an islanded microgrid in rural area consisting of PV, wind, biomass and battery energy storage system," *Energy Conversion and Management*, vol. 128, pp. 178–190, 2016.

[22] J. Li, P. Liu, and Z. Li, "Optimal design and techno-economic analysis of a solar-wind-biomass off-grid hybrid power system for remote rural electrification: A case study of west China," *Energy*, vol. 208, p. 118387, 2020.

[23] B. K. Das, N. Hoque, S. Mandal, T. K. Pal, and M. A. Raihan, "A techno-economic feasibility of a stand-alone hybrid power generation for remote area application in Bangladesh," *Energy*, vol. 134, pp. 775–788, 2017.

[24] J. Ahmad, M. Imran, A. Khalid, *et al.*, "Techno economic analysis of a wind-photovoltaic-biomass hybrid renewable energy system for rural electrification: A case study of Kallar Kahar," *Energy*, vol. 148, pp. 208–234, 2018.

[25] L. Lozano, E. M. Querikiol, M. L. S. Abundo, and L. M. Bellotindos, "Techno-economic analysis of a cost-effective power generation system for off-grid island communities: A case study of Gilutongan Island, Cordova, Cebu, Philippines," *Renewable Energy*, vol. 140, pp. 905–911, 2019.

[26] L. M. Halabi and S. Mekhilef, "Flexible hybrid renewable energy system design for a typical remote village located in tropical climate," *Journal of Cleaner Production*, vol. 177, pp. 908–924, 2018.

[27] F. Akram, F. Asghar, M. A. Majeed, W. Amjad, M. O. Manzoor, and A. Munir, "Techno-economic optimization analysis of stand-alone renewable energy system for remote areas," *Sustainable Energy Technologies and Assessments*, vol. 38, p. 100673, 2020.

[28] M. L. Kolhe, K. I. U. Ranaweera, and A. S. Gunawardana, "Techno-economic sizing of off-grid hybrid renewable energy system for rural electrification in Sri Lanka," *Sustainable Energy Technologies and Assessments*, vol. 11, pp. 53–64, 2015.

[29] A. Asrari, A. Ghasemi, and M. H. Javidi, "Economic evaluation of hybrid renewable energy systems for rural electrification in Iran—A case study," *Renewable and Sustainable Energy Reviews*, vol. 16, no. 5, pp. 3123–3130, 2012.

[30] A. H. Mamaghani, S. A. A. Escandon, B. Najafi, A. Shirazi, and F. Rinaldi, "Techno-economic feasibility of photovoltaic, wind, diesel and hybrid electrification systems for off-grid rural electrification in Colombia," *Renewable Energy*, vol. 97, pp. 293–305, 2016.

[31] K. Gebrehiwot, M. A. H. Mondal, C. Ringler, and A. G. Gebremeskel, "Optimization and cost-benefit assessment of hybrid power systems for off-grid rural electrification in Ethiopia," *Energy*, vol. 177, pp. 234–246, 2019.

[32] M. Ramesh and R. P. Saini, "Dispatch strategies based performance analysis of a hybrid renewable energy system for a remote rural area in India," *Journal of Cleaner Production*, vol. 259, p. 120697, 2020.

[33] A. Bhatt, M. Sharma, and R. Saini, "Feasibility and sensitivity analysis of an off-grid micro hydro–photovoltaic–biomass and biogas–diesel–battery

hybrid energy system for a remote area in Uttarakhand state, India," *Renewable and Sustainable Energy Reviews*, vol. 61, pp. 53–69, 2016.

[34] A. K. Nag and S. Sarkar, "Modeling of hybrid energy system for futuristic energy demand of an Indian rural area and their optimal and sensitivity analysis," *Renewable Energy*, vol. 118, pp. 477–488, 2018.

[35] O. D. T. Odou, R. Bhandari, and R. Adamou, "Hybrid off-grid renewable power system for sustainable rural electrification in Benin," *Renewable Energy*, vol. 145, pp. 1266–1279, 2020.

[36] A. A. Z. Diab, H. M. Sultan, I. S. Mohamed, O. N. Kuznetsov, and T. D. Do, "Application of different optimization algorithms for optimal sizing of PV/ wind/diesel/battery storage stand-alone hybrid microgrid," *IEEE Access*, vol. 7, pp. 119223–119245, 2019.

[37] A. A. Z. Diab, H. M. Sultan, and O. N. Kuznetsov, "Optimal sizing of hybrid solar/wind/hydroelectric pumped storage energy system in Egypt based on different meta-heuristic techniques," *Environmental Science and Pollution Research*, vol. 27, no. 26, pp. 32318–32340, 2020.

[38] D. H. Wolpert and W. G. Macready, "No free lunch theorems for optimization," *IEEE Transactions on Evolutionary Computation*, vol. 1, no. 1, pp. 67–82, 1997.

[39] A. A. Z. Diab, H. M. Sultan, T. D. Do, O. M. Kamel, and M. A. Mossa, "Coyote optimization algorithm for parameters estimation of various models of solar cells and PV modules," *IEEE Access*, vol. 8, pp. 111102–111140, 2020.

[40] H. M. Sultan, A. A. Z. Diab, N. K. Oleg, and S. Z. Irina, "Design and evaluation of PV-wind hybrid system with hydroelectric pumped storage on the National Power System of Egypt," *Global Energy Interconnection*, vol. 1, no. 3, pp. 301–311, 2018.

[41] H. M. Sultan, O. N. Kuznetsov, A. S. Menesy, and S. Kamel, "Optimal configuration of a grid-connected hybrid PV/wind/hydro-pumped storage power system based on a novel optimization algorithm," in *2020 International Youth Conference on Radio Electronics, Electrical and Power Engineering (REEPE)*, 2020, pp. 1–7: IEEE.

[42] O. Erdinc and M. Uzunoglu, "Optimum design of hybrid renewable energy systems: Overview of different approaches," *Renewable and Sustainable Energy Reviews*, vol. 16, no. 3, pp. 1412–1425, 2012.

[43] L. Olatomiwa, S. Mekhilef, M. S. Ismail, and M. Moghavvemi, "Energy management strategies in hybrid renewable energy systems: A review," *Renewable and Sustainable Energy Reviews*, vol. 62, pp. 821–835, 2016.

[44] F. Bourennani, S. Rahnamayan, and G. Naterer, "Optimal design methods for hybrid renewable energy systems," *International Journal of Green Energy*, vol. 12, no. 2, pp. 148–159, 2015.

[45] J. Jung and M. Villaran, "Optimal planning and design of hybrid renewable energy systems for microgrids," *Renewable and Sustainable Energy Reviews*, vol. 75, pp. 180–191, 2017.

[46] H. Abd El-Sattar, S. Kamel, M. A. Tawfik, D. Vera, and F. Jurado, "Modeling and simulation of corn stover gasifier and micro-turbine for power generation," *Waste and Biomass Valorization*, vol. 10, pp. 3101–3114, 2019.

[47] H. Abd El-Sattar, H. M. Sultan, S. Kamel, T. Khurshaid, and C. Rahmann, "Optimal design of stand-alone hybrid PV/wind/biomass/battery energy storage system in Abu-Monqar, Egypt," *Journal of Energy Storage*, vol. 44, p. 103336, 2021.

[48] M. Eteiba, S. Barakat, M. Samy, and W. I. Wahba, "Optimization of an off-grid PV/Biomass hybrid system with different battery technologies," *Sustainable Cities and Society*, vol. 40, pp. 713–727, 2018.

[49] A. Cano, P. Arévalo, and F. Jurado, "Energy analysis and techno-economic assessment of a hybrid PV/HKT/BAT system using biomass gasifier: Cuenca-Ecuador case study," *Energy*, vol. 202, p. 117727, 2020.

[50] M. Samy, H. I. Elkholy, and S. Barakat, "Multi-objective optimization of hybrid renewable energy system based on biomass and fuel cells," *International Journal of Energy Research*, vol. 45, no. 6, pp. 8214–8230, 2021.

[51] S. Vendoti, M. Muralidhar, and R. Kiranmayi, "Techno-economic analysis of off-grid solar/wind/biogas/biomass/fuel cell/battery system for electrification in a cluster of villages by HOMER software," *Environment, Development and Sustainability*, vol. 23, no. 1, pp. 351–372, 2021.

[52] H. R. Baghaee, M. Mirsalim, G. B. Gharehpetian, and H. Talebi, "Reliability/cost-based multi-objective Pareto optimal design of stand-alone wind/PV/FC generation microgrid system," *Energy*, vol. 115, pp. 1022–1041, 2016.

[53] H. M. Sultan, A. S. Menesy, S. Kamel, A. Korashy, S. Almohaimeed, and M. Abdel-Akher, "An improved artificial ecosystem optimization algorithm for optimal configuration of a hybrid PV/WT/FC energy system," *Alexandria Engineering Journal*, vol. 60, no. 1, pp. 1001–1025, 2021.

[54] S. A. N'guessan, K. S. Agbli, S. Fofana, and D. Hissel, "Optimal sizing of a wind, fuel cell, electrolyzer, battery and supercapacitor system for off-grid applications," *International Journal of Hydrogen Energy*, vol. 45, no. 8, pp. 5512–5525, 2020.

[55] F. A. Alturki and E. M. Awwad, "Sizing and cost minimization of standalone hybrid wt/pv/biomass/pump-hydro storage-based energy systems," *Energies*, vol. 14, no. 2, p. 489, 2021.

[56] A. Qi, D. Zhao, A. A. Heidari, L. Liu, Y. Chen, and H. Chen, "FATA: an efficient optimization method based on geophysics," *Neurocomputing*, vol. 607, p. 128289, 2024.

Chapter 5

Optimization and sizing of isolated hybrid solar PV/DG/battery energy systems for residential community in hot climate areas of Saudi Arabia

Ahmed S. Menesy[1], Hamdy M. Sultan[2],
Ibrahim O. Habiballah[1], Mahmoud Kassas[1,3] and
Salah Kamel[4]

Abstract

This chapter focuses on the development of a smart power system that integrates renewable energy resources to optimize the supply of clean electricity, particularly for rural areas in hot climates. The main objective is to reduce energy production costs (EC) while enhancing the reliability and efficiency of power delivery in remote rural areas. To achieve this, this chapter explores two advanced optimization techniques, namely, the hippopotamus optimization (HO) and the greylag goose optimization (GGO) techniques. These techniques are applied to a hybrid energy system configuration that combines solar photovoltaic (PV) panels, diesel generators (DG), and battery energy storage to meet energy demands with increased stability and efficiency. Through comparative analysis, this study reveals that the HO algorithm demonstrates greater efficiency than the GGO, showing a shorter runtime and more robust convergence characteristics. Simulation results indicate that the HO algorithm achieves a superior best fitness value of 0.0963 compared to GGO's 0.1129, highlighting HO's better convergence behavior and efficiency. Additionally, the HO-optimized system shows a net present cost (NPC) of 21.3 M$ and a cost of energy (COE) of 0.33 $/kWh, which are significantly lower than GGO's 27.2 M$ NPC and 0.42 $/kWh COE. In terms of energy reliability, GGO achieved a lower loss of power supply probability (LPSP) of 0.022 compared to HO's 0.0367, indicating its strength in maintaining a continuous power supply. Meanwhile,

[1]Electrical Engineering Department, King Fahd University of Petroleum and Minerals, Saudi Arabia
[2]Electrical Engineering Department, Faculty of Engineering, Minia University, Egypt
[3]Interdisciplinary Research Center for Sustainable Energy Systems, King Fahd University of Petroleum and Minerals, Saudi Arabia
[4]Department of Electrical Engineering, Faculty of Engineering, Aswan University, Egypt

the HO approach reduces the excess energy directed to the dummy load to 0.0398, offering improved economic efficiency. These results underscore the potential for the proposed hybrid PV/DG/battery system to deliver cost-effective, stable energy for off-grid communities, offering sustainable energy solutions with minimal reliance on conventional power sources. Furthermore, this chapter provides insights into optimization strategies that can be applied in smart power systems to facilitate sustainable energy solutions in off-grid and rural regions

Keywords: Hybrid microgrid; Renewable energy sources; Optimal sizing; PV; Battery; LPSP; Optimization

5.1   Introduction

In recent years, fossil fuels like coal, natural gas, and oil have been the primary sources of global electricity production. However, these resources raise concerns due to their greenhouse gas emissions, which accelerate climate change, and their finite availability, which results in price volatility and future scarcity [1,2]. The surge in worldwide energy consumption poses a substantial challenge to existing power systems, emphasizing the necessity of adopting renewable energy solutions to address growing energy demands [3,4]. Despite the optimistic trajectory, renewable energy systems face inherent limitations that challenge their broader adoption [5,6]. These include variability in power generation due to dependency on environmental conditions, limited storage capabilities, and higher initial capital costs compared to traditional fossil fuel-based systems [7,8]. Rising electricity demand and growing public and regulatory calls for emissions reduction highlight the urgent need for sustainable, environmentally friendly energy alternatives. Renewable energy sources (RESs), like solar, wind, hydropower, biomass, ocean, and geothermal power, offer a promising solution. These clean energy sources produce minimal greenhouse gas emissions and have a far smaller environmental impact than fossil fuels. Additionally, renewable resources are virtually limitless, ensuring a reliable energy supply. Microgrids represent a significant innovation in the energy sector, operating independently or in conjunction with traditional grids. They integrate various renewable sources, enhancing the resilience and sustainability of energy infrastructure. Microgrids support the clean energy transition by reducing fossil fuel dependence and fostering an eco-friendly energy system. This transition toward renewable energy is gaining momentum worldwide, supported by commitments made at COP26 in Glasgow, where many nations pledged to meet ambitious emissions targets by 2030 and achieve net zero by mid-century. Key drivers of this shift include energy security from domestically generated renewables and the economic competitiveness of solar and wind energy amid high fossil fuel prices. In 2021, global renewable energy capacity reached 3064 GW, with hydropower at 1230 GW, solar at 849 GW, and wind at 825 GW, among other sources [9]. Saudi Arabia exemplifies this commitment with its Vision 2030 goals to achieve net zero by 2060 and produce 50% of its electricity from RES by

2030, planning to install 58.7 GW of RES capacity, including 40 GW from solar and 16 GW from wind.

Among the diverse renewable options, solar PV, wind, and geothermal stand out for their potential and applicability [10,11]. Solar PV, in particular, has gained traction due to significant cost reductions. Despite their benefits, these RESs share a challenge: their output is less predictable and controllable than fossil-fuel-based generation. Fossil fuel generators offer a stable and adjustable power supply that renewable sources, dependent on natural conditions like sunlight and wind, cannot match. This variability complicates their integration into power grids, necessitating advancements in power electronics to manage and stabilize renewable output [12,13]. Hybrid microgrid systems have emerged as a robust solution, integrating renewable resources with battery storage and diesel generators (DG) to manage variability and maintain reliability. Hybrid renewable energy systems (HRES) in microgrids can operate either as stand-alone systems or be integrated into the main grid. Grid-connected systems rely on the main grid for backup, ensuring reliability, while stand-alone systems require careful sizing of storage and generation capacity to meet demand reliably. Oversizing improves reliability but increases costs, whereas undersizing lowers costs but risks power shortages. An optimal balance, typically measured by a reliability index, helps design HRESs that meet demand without excessive costs. Hybrid microgrid systems combine renewables like solar PV and wind with energy storage and traditional backup, such as batteries and diesel generators, providing stable and reliable power. Storage systems can hold excess energy during high production, while diesel generators offer backup as needed, ensuring a continuous supply. By leveraging renewable, storage, and backup sources, hybrid microgrids effectively manage renewable variability, providing a steady and dependable energy supply [13,14].

Metaheuristic algorithms, inspired by natural processes, have recently become key in optimizing microgrid component sizing due to their ability to navigate large solution spaces and avoid local optima. Algorithms like genetic algorithms, particle swarm optimization, and simulated annealing are particularly effective for complex, multi-dimensional problems where flexibility and robustness are essential [15]. While software tools simplify straightforward analyses and deterministic methods offer precise solutions for well-defined tasks, metaheuristics excel in challenging optimization scenarios. This range of approaches enables researchers and engineers to optimize hybrid RES configurations, balancing reliability and other critical factors effectively. In a case study in Arar city, Saudi Arabia, Eltamaly et al. [16] proposed an autonomous HRES, including WTs, PV panels, and batteries, to supply an average of 1000 m^3/day. The hybrid system has been developed to produce 2440 kW to meet water production needs at minimal cost and with minimal supply interruption. A matching study was conducted to identify the best WT among 10 market-available options. Three optimization techniques—particle swarm optimization, bat algorithm (BA), and social mimic optimization—were compared to avoid premature convergence. Results demonstrated the superiority of the RES in powering the reverse osmosis desalination (ROD) plant, with the BA being the fastest and most accurate approach with a cost of water of $0.745/m^3. Ali et al. [17] provided an accurate model that relies on an energy management strategy (EMS) to power small-scale

ROD systems in off-grid communities with limited access to freshwater. The main results of their study have revealed a substantial interconnection between pump capacity and power/water control. In a similar manner, an economic model was explored for isolated and grid-dependent hybrid PV/WT/battery systems in a certain area in China [18]. The findings indicated that wind energy alone may not be sufficient to meet the load requirements in Guiyang. Furthermore, Aziz *et al.* [19] introduced a dispatch strategy, implemented through HOMER software, to address the drawbacks of HOMER scenarios for hybrid energy systems. The results highlight that the suggested approach demonstrates superior economic and environmental performance, boasting a net present cost (NPC) of $56,473 and annual CO_2 emissions totaling 6,838 kg. Additionally, a sensitivity analysis underscores the robustness of the proposed strategy, indicating minimal susceptibility to fluctuations in fuel prices. This stands in stark contrast to the load following (LF) and cycle charging (CC) strategies, which exhibit pronounced vulnerability to such price variations, thereby reinforcing the resilience and stability of the proposed methodology. These results suggest that the proposed strategy is a more realistic and efficient energy management approach, particularly in regions with volatile fuel prices.

Ajiwiguna *et al.* [20] suggested a methodology to optimize the sizes of hybrid system components that incorporate PV-RO technology along with water storage to minimize overall costs. Their results conclusively demonstrated that incorporating a seasonal water storage tank (SWST) in PV-powered water systems effectively stabilizes water production without the need for costly batteries. Furthermore, operating the RO unit for 8–9 h per day results in highly competitive water costs, establishing this approach as a practical and cost-effective solution for ensuring consistent and reliable water supply. Khiari *et al.* [21] proposed a power management methodology for off-grid hybrid RES including PV/WT, without the need for battery storage to meet the electricity demand of a brackish water ROD plant. The results concluded that the novel power control strategy that uses power field-oriented control successfully operates a brackish water reverse osmosis desalination test bench with isolated hybrid RES including PV and wind, eliminating the need for batteries, while maintaining stable DC bus voltage and consistent power transfer under varying generation conditions. The strategy's effectiveness was verified through experimental testing across different operating modes while adhering to a safe operating window (SOW) for desalination, and the inclusion of a basic energy management system (EMS) ensured reliable power sharing. Elmaadawy *et al.* [22] used HOMER software to achieve the optimal configuration of an isolated hybrid system encompassing PV, WT, DG, battery storage, and a converter to meet the power requirements of a bulk ROD plant. Their research emphasized the indispensability of a battery storage unit to supply power during nighttime. Mehrjerdi [23] introduced a hybrid scheme combining solar, wind, and batteries with three distinct desalination methods, namely multi-stage flash, RO, and multi-effect distillation. The study established that the hybrid power system employing RO desalination emerged as the best economically viable option.

The main objectives of this chapter are to develop two recent optimization algorithms, namely, the hippopotamus optimization (HO) [24] and greylag goose optimization (GGO) [25], for the optimal design of stand-alone hybrid systems, specifically

PV/DG/battery energy systems. Additionally, the chapter aims to conduct a cost-effective analysis to showcase the economic viability and potential for broad adoption of the proposed hybrid system.

The organization of this chapter is as follows: Section 5.2 presents the mathematical modeling and description of the proposed hybrid system, detailing the system components and their interactions. Section 5.3 describes the EMS employed to optimize the system's performance. Section 5.4 defines the objective function and constraints used in the optimization process. Section 5.5 introduces the HO and GGO algorithms and explains their implementation. Section 5.6 provides the simulation results and analysis, offering a comprehensive evaluation of the system's performance. Finally, Section 5.7 concludes the chapter with a summary of the findings, highlighting the key contributions and potential future research directions.

5.2   Proposed hybrid system description and mathematical modeling

The hybrid microgrid energy systems offer an ideal solution for addressing electrification challenges in islands and remote locations, where extending the transmission network is often neither economical nor feasible. These hybrid systems integrate multiple RESs, such as solar (PV) and wind, alongside battery storage and DGs, to create a reliable and sustainable power supply. The specific hybrid microgrid configuration being studied includes four primary components: PV panels, a DG, an inverter, and a battery storage unit. For remote regions, a PV/diesel/battery hybrid system can provide a cost-effective approach to meet electricity demands, reducing reliance on conventional energy sources while enhancing energy security. By combining the strengths of each component, the system ensures a steady power supply despite variations in renewable generation. Figure 5.1 illustrates the schematic

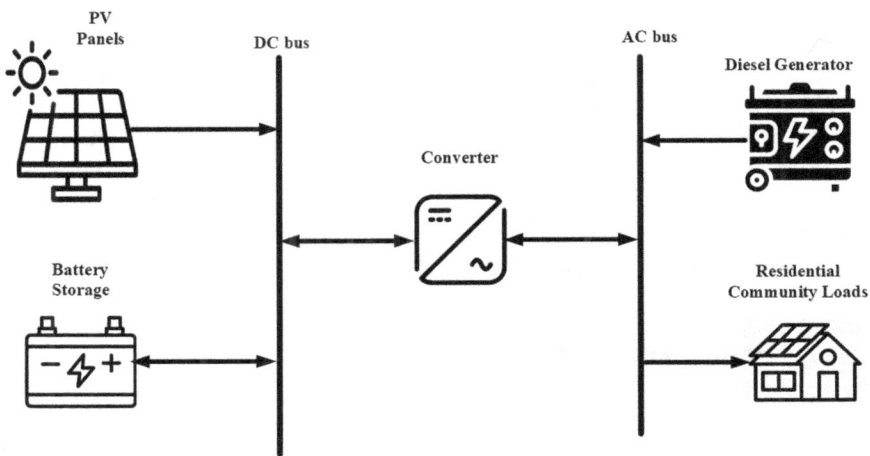

Figure 5.1   Schematic diagram of the proposed hybrid renewable energy system

presentation of this suggested hybrid microgrid system, highlighting the integration and interaction between the various components.

5.2.1 System modeling

5.2.1.1 Solar PV system

The energy output of a PV system is affected by several factors, including the intensity of solar irradiation at the location and the ambient temperature surrounding the PV modules. This relationship can be expressed in the following form [26,27]:

$$P_{PV}(t) = A_{PV} P_{PV} \eta_{PV} \eta_{INV} \eta_{Wire} \frac{G(t)}{G_{nom}} (1 - \beta_T(T_C(t) - T_{C_{nom}})) \tag{5.1}$$

where A_{PV} represents the area of the PV module, P_{PV} is the installed capacity of each PV module, η_{PV} is the conversion efficiency of the PV module, η_{INV} denotes the inverter efficiency, η_{Wire} accounts for the efficiency losses in wiring, G is the actual solar irradiation intensity incident on the PV module, G_{nom} is the solar irradiation intensity under standard test conditions (STC), β_T is the temperature coefficient of power for the module, which indicates how the module's power output varies with temperature, T_C is the temperature of the solar cell under operational conditions, and $T_{C_{nom}}$ is the cell temperature under STC.

The temperature of the PV cell, T_C, is crucial in determining the system's performance and can be estimated based on the surrounding air temperature as follows [28]:

$$T_C(t) = T_{ambient} + G(t) \frac{T_{Test}}{800} \tag{5.2}$$

where $T_{ambient}$ signifies the ambient temperature and T_{Test} represents a reference temperature related to the component's thermal response. In this model, ambient factors, especially temperature and solar radiation, play a crucial role in determining the energy yield of the system. Typically, as solar cell heats up, the efficiency of the PV module decreases. β_T indicates how sensitive the power output of the PV module is to temperature variations, which is valuable for evaluating potential efficiency losses in warmer climates. By integrating these parameters, the PV system model provides a realistic estimation of energy production across different environmental conditions, including varying levels of irradiance and temperature. This capability enables system designers and energy planners to optimize PV installations for specific locations, ensuring they effectively meet energy demands.

5.2.1.2 Battery energy storage system

The incorporation of RES, such as solar and wind, has significantly altered the energy landscape. Nevertheless, the inherent irregularity and intermittency of these sources can result in variations in power generation. For example, solar energy output is highest on sunny days but decreases considerably at night or during overcast conditions. Likewise, wind energy generation fluctuates according to changing wind patterns. To maintain a consistent and dependable power supply that meets load requirements, a robust energy storage system is crucial. Energy storage serves to reconcile the disparity between energy production and consumption, enabling the capture of surplus energy generated during peak periods for use during times of low generation [29,30].

In this study, lead-acid batteries are identified as the primary storage solution. These batteries offer several advantages, such as low initial costs, widespread availability, and proven technology. However, they also present certain drawbacks, including a shorter lifespan when compared to more advanced technologies like lithium-ion batteries, which provide higher energy densities and longer cycle lives. Recognizing these distinctions is important for choosing the most suitable battery technology based on the specific requirements of the project and budgetary considerations [31,32].

To evaluate the capacity of the battery system, the state of charge (SOC) is commonly used as an indicator. The SOC reflects the current energy level within the battery system and can be determined through continuous monitoring of the charging and discharging processes. Depth of discharge (DOD) represents the percentage of the battery's total capacity, which was depleted in relation to its overall capacity [33,34]. A lower DOD can extend the battery's lifespan, whereas a higher DOD may result in faster deterioration. The calculation of the SOC at any given time can be expressed as follows [33,35]:

Charging phase

$$E_{CH} = \left(\frac{-P_{load}(t)}{\eta_{INV}} + P_{PV}(t) \right) . \eta_{CH} \tag{5.3}$$

$$SOC(t) = SOC(t-1)(1-\sigma) + E_{CH} \tag{5.4}$$

Discharging phase

$$E_{DIS} = \left(\frac{P_{load}(t)}{\eta_{INV}} - P_{PV}(t) \right) . \eta_{DIS} \tag{5.5}$$

$$SOC(t) = SOC(t-1)(1-\sigma) - E_{DIS} \tag{5.6}$$

In these equations, $OC(t)$ represents the SOC at time t, and $SOC(t-1)$ denotes the state of charge from the previous time step. The parameter σ signifies the self-discharge rate of the battery. The terms P_{PV} and P_{load} correspond to the energy produced by renewable sources and the energy demand, respectively. Additionally, η_{INV}, η_{CH}, and η_{DIS} represent the efficiency of the inverter, the charging efficiency, and the discharging efficiency of the battery storage system. Incorporating a thoughtfully designed battery storage system is essential for improving the reliability and stability of renewable energy systems. By efficiently managing the SOC and taking into account the different factors that affect battery performance, this strategy can greatly enhance energy management practices in renewable energy applications.

5.2.1.3   Diesel generator

DG is a traditional source of electric energy commonly utilized in hybrid power systems as a supplementary power source. DG supplies energy when the combined output from the PV and wind systems is insufficient to meet energy demands, especially when the battery bank has been exhausted. In such scenarios, the DG kicks in to provide the necessary power. Integrating a DG into a hybrid power system provides a dependable power supply during times of reduced renewable generation [36,37]. This functionality is crucial for maintaining system stability and fulfilling peak load requirements. However, dependence on DGs brings forth issues related to fuel expenses, emissions, and sustainability, emphasizing the importance of efficient management and integration within the overall energy framework. Therefore, optimizing the operation of the DG alongside RESs and energy storage can create a more balanced and eco-friendly energy solution. The fuel consumption per hour of the DG, denoted as $D_f(t)$, can be calculated using the following equation [38]:

$$D_f(t) = \alpha_{DG} P_{DG}(t) + \beta_{DG} P_{DG_R} \tag{5.7}$$

In this equation, $P_{DG}(t)$ represents the average power output of the DG over the hour, P_{DG_R} implies the rated power of the DG, α_{DG} and β_{DG} are coefficients derived from the fuel consumption curve of the DG. For this study, the coefficients have been assigned values of 0.246 and 0.08145 for α_{DG} and β_{DG}, respectively [35,38].

Key specifications of the technical parameters of all components for the proposed HRES are tabulated in Table 5.1.

5.2.2   Energy management strategy

When the hybrid energy system operates in an isolated mode and the power produced by the PV system falls short of the energy demand from the load, the storage devices will enter discharge mode to supply the needed energy. The relationship between load demand, power from the PV system, and the power supplied by the battery can be expressed as

$$P_{load}(t) = (P_{PV}(t) + P_{DIS}(t)) * \eta_{INV} \tag{5.8}$$

In this scenario, the battery compensates for the gap in power generation, ensuring that the load demand is met. If the combined power from the PV system and the battery discharge still does not meet the energy demand, the deficit will be supplied by a DG. The power contribution from the DG can be formulated using

$$P_{DG}(t) = P_{load}(t) - (P_{PV}(t) + P_{DIS}(t)) * \eta_{INV} \tag{5.9}$$

This equation ensures that the total demand is satisfied by accounting for any shortfall after considering the contributions from the PV system and the battery.

Conversely, when the power supplied by the PV system exceeds the load requirement, the surplus energy will be utilized to charge the battery. The power

Table 5.1   Technical and economic attributes of the system components under study

Component	Parameter	Value	Unit
Solar PV [39]	Model	LR5-54HPH-420M	
	Unit power	260	watt
	Length	1625	mm
	Width	1019	mm
	Thickness	46	mm
	Module efficiency	15.7	%
	Operating temperature	47	°C
	Temperature coefficient	0.45	%
	Initial cost	112	$
	O&M cost	1	% of initial cost
	Lifetime	25	year
Battery bank [35]	Model	RS lead acid battery	
	Capacity	50	Ah
	Voltage	12	V
	efficiency	86	%
	DOD	80	%
	weight	16.5	kg
	Max. discharge current	750	A
	Internal resistance	<=0.006	Ω
	Operating temperature	0-40	°C
	Initial cost	146.5	$
	Replacement cost	102.55	$
	Lifetime	10	year
DG [35]	Rated power	100	kW
	Initial cost	850	$/kW
	Replacement cost	850	$/kW
	O&M cost	3	% of initial cost
	Lifetime	10	year
Inverter [35]	Rated power	1	kW
	Initial cost	711	$/kW
	Replacement cost	650	$/kW
	O&M cost	3	% of initial cost
	Lifetime	10	year
	efficiency	95	%

flow in this scenario can be expressed as

$$P_{load}(t) = (P_{PV}(t) - P_{CH}(t)) * \eta_{INV} \qquad (5.10)$$

Here, the equation indicates that the load is satisfied by the power supplied by PV panels, minus the amount diverted to charge the battery. The inverter's efficiency is again factored in, reflecting the real power available to the load after charging.

When the battery banks get fully charged, any additional power generated by the PV system will be diverted to a dummy load instead of charging the battery further. This is expressed as

$$P_{dummy}(t) = P_{load}(t) + (P_{PV}(t) - P_{CH}(t)) * \eta_{INV} \qquad (5.11)$$

This equation indicates that the excess power generated by the PV system, which is not used for charging, is redirected to the dummy load to prevent over-charging and damage to the battery.

Figure 5.2 presents a flowchart that depicts the operational logic of the isolated hybrid energy system. This visual tool highlights the interactions among various components, such as the PV modules, battery, DG, load, and dummy load. It serves to guide users through the decision-making processes associated with managing energy production and consumption effectively.

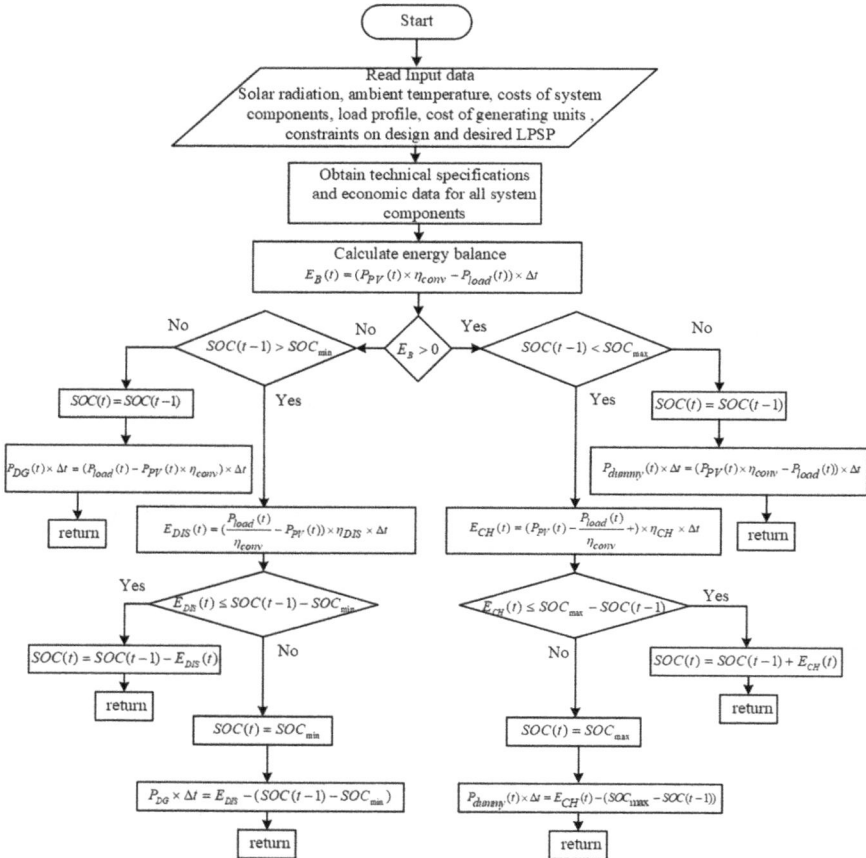

Figure 5.2   Flowchart illustrating the proposed stand-alone hybrid microgrid system

5.3 Optimization problem formulation

5.3.1 Optimization problem parameters' evaluation

This study focuses on evaluating three key parameters as objective functions: the cost of energy (COE), the loss of power supply probability (LPSP), and excess power dissipated in dumper load (EXP). The primary goal is to optimize the capacities of an HRES to guarantee the reliability of the energy solution while reducing costs.

5.3.1.1 Cost of energy

To assess the total cost of the hybrid microgrid, the NPC is utilized. The annual investment cost of the system, denoted as C_{ann_tot}, is formulated as

$$C_{ann\_tot} = C_{ann\_cap} + C_{ann\_rep} + C_{ann\_O\&M} \tag{5.12}$$

where C_{ann_cap} denotes the annual investment cost of each subsystem, C_{ann_rep} denotes the annual cost for replacements of each subsystem, and $C_{ann\_O\&M}$ denotes the annual operating and maintenance cost.

(a) *The capital cost:* The capital recovery factor (CRF) is used to translate the initial investment cost into an equivalent annual capital expense. The formula for CRF is given by

$$\mathrm{CRF}(r,n) = \frac{r(1+r)^n}{(1+r)^n - 1} \tag{5.13}$$

where r denotes the interest rate (as a percentage) and n denotes project life-span (in years).

The annual capital expense for each sub-system is formulated utilizing the following equations:

$$\begin{cases} C_{ann\_PV} = C_{cap\_PV} * \mathrm{CRF}(r, M_{PV}) \\ C_{ann\_batt} = C_{cap\_batt} * \mathrm{CRF}(r, M_{batt}) \\ C_{ann\_DG} = C_{cap\_DG} * \mathrm{CRF}(r, M_{DG}) \\ C_{ann\_INV} = C_{cap\_INV} * \mathrm{CRF}(r, M_{INV}) \end{cases} \tag{5.14}$$

where C_{cap_PV}, C_{cap_batt}, C_{cap_DG}, and C_{cap_INV} represent the initial capital costs for the installation of the PV, battery, DG, and inverter. M_{PV}, M_{batt}, M_{DG}, and M_{INV} denote the lifespan of the PV modules, battery banks, DGs, and inverters, respectively.

The total annual capital investment for the system is calculated as

$$C_{ann\_cap} = C_{ann\_PV} + C_{ann\_DG} + C_{ann\_batt} + C_{ann\_INV} \tag{5.15}$$

where $C_{ann\_cap\_PV}$, $C_{ann\_cap\_DG}$, $C_{ann\_cap\_batt}$, and $C_{ann\_cap\_INV}$ represent the annual shares of the capital costs for the PV system, DG, battery bank, and inverter.

(b) *The O&M cost:* The operating and maintenance costs for the hybrid system are calculated as follows:

$$C_{ann\_O\&M} = C_{O\&M\_PV} + C_{O\&M\_DG} + C_{O\&M\_batt} + C_{O\&M\_INV} \qquad (5.16)$$

where, $C_{O\&M\_PV}$, $C_{O\&M\_DG}$, $C_{O\&M\_batt}$, and $C_{O\&M\_INV}$ are the operating and maintenance costs associated with the PV system, DG, battery banks, and inverter, respectively.

(c) *The annual replacement cost:* The present value of the replacement costs for components within the hybrid system over its lifespan can be calculated using the following equation [40]:

$$C_{ann\_rep} = C_{rep\_PV} + C_{rep\_DG} + C_{rep\_batt} + C_{rep\_INV} \qquad (5.17)$$

where $C_{O\&M\_PV}$, $C_{O\&M\_DG}$, $C_{O\&M\_batt}$, and $C_{O\&M\_INV}$ are the operating and maintenance costs associated with the PV system, DG, battery banks, and inverter, respectively.

The NPC of the hybrid system is computed using the following formula:

$$\text{NPC} = \frac{C_{ann\_tot}}{\text{CRF}} \qquad (5.18)$$

The COE, which indicates the cost of the electrical energy produced by the suggested microgrid (in \$/kWh), is expressed as follows:

$$\text{COE} = \frac{C_{ann\_tot}}{\sum_{h=1}^{h=8760} P_{load}} = \frac{\text{NPC}}{\sum_{h=1}^{h=8760} P_{load}} * \text{CRF} \qquad (5.19)$$

where P_{load} represents the value of hourly demand over the year.

5.3.1.2 Loss of power supply probability

The LPSP is a critical design metric that quantifies the likelihood of inadequate power delivery when the hybrid microgrid fails to meet energy demand. This situation arises when the total generation falls short of the load requirements. The loss of power supply (LPS) at any given moment, denoted as LPS(t), can be expressed mathematically as

$$
\begin{aligned}
\text{LPS}(t) \quad &= P_{Load}(t).\Delta t \\
&- (P_{DG}(t)).\Delta t \\
&- ((P_{PV}(t).\Delta t) + SOC(t - \Delta t) - SOC_{\min}
\end{aligned}
\qquad (5.20)
$$

LPSP serves as a crucial factor for assessing the reliability of HRESs, especially in optimal sizing problems. It provides insights into the system's ability to consistently deliver power to meet demand. The overall LPSP can be calculated by aggregating all observed values of LPS(t) across the entire study period:

$$\text{LPSP} = \frac{\sum_{t=1}^{8760} \text{LPS}(t)}{\sum_{t=1}^{8760} P_{Load}(t).\Delta t} \tag{5.21}$$

This formulation yields a ratio that reflects the frequency and severity of supply shortages relative to total energy demands. Furthermore, during the reliability analysis of the system, the following condition must hold:

$$P_{Load}(t) > P_{tot\_generation}(t) \tag{5.22}$$

This requirement underscores the importance of having the total generation capacity—encompassing all sources, such as PV systems and DGs—surpass the load demand at any moment to prevent power supply disruptions. By consistently monitoring and evaluating these variables, engineers can guarantee that the hybrid microgrid is properly sized and capable of accommodating variable energy demands, thereby improving the overall reliability of the system.

5.3.1.3 Excess energy in dumper load

In hybrid systems, particularly those including RESs, it is common to encounter scenarios where the power generated exceeds the current load consumption. When this occurs, the surplus energy must be managed effectively to prevent issues like the overcharging of energy storage systems or instability in the power supply. One effective solution is the use of a dummy load, which can absorb this excess energy.

The amount of energy directed to the dummy load can be calculated based on the difference between the total energy produced by the generation sources and the energy required by the load. The formula can be expressed as

$$P_{dum}(t) = P_{tot_{generation}}(t) - P_{Load}(t) \tag{5.23}$$

The excess energy absorbed is expressed as

$$P_{EXC} = \sum_{1}^{8760} (P_{dum}/P_{Load}) \tag{5.24}$$

5.3.2 Problem constraints

In off-grid hybrid systems, various components operate under specific constraints to ensure optimal performance. To maintain balance in the power supply, the system must satisfy the constraint function outlined in (5.25), which asserts that the sum of the generation from all sources must equal the load demand at any given time

$$P_{PV}(t) + P_{DG}(t) + P_{batt}(t) = P_{Load}(t) \tag{5.25}$$

To effectively manage the output of the DG, the optimization strategy ensures that the power produced by the DG does not exceed its rated capacity, as defined in the following equation:

$$P_{DG}(t) \leq P_{DG\_rated} \tag{5.26}$$

Additionally, to prevent issues related to overcharging or undercharging the battery storage system, the SOC must adhere to design limits throughout the operational period. The SOC must be maintained within the parameters specified in equation (5.27). The maximum SOC, denoted as SOC_{max}, represents the full capacity of the battery, while the minimum SOC, SOC_{min}, is primarily influenced by the battery's depth of discharge

$$SOC_{max} \leq SOC(t) \leq SOC_{min} \tag{5.27}$$

Furthermore, the LPSP must remain below a predefined reliability threshold LPSP, as expressed in (5.28). In this study, it is stipulated that β_L should not exceed 0.06

$$LPSP \leq \beta_L \tag{5.28}$$

By complying with these constraints, the hybrid system can function optimally, guaranteeing dependable energy supply while extending the longevity of its components and promoting sustainability. This structure facilitates the efficient management of energy resources, especially in situations where supply and demand fluctuate, thereby enhancing the system's reliability and overall performance.

5.3.3   Objective function

The goal of the optimization issue is to determine the ideal configuration of the system components to maximize the energy output of the HRES while simultaneously minimizing the energy cost (EC). Additionally, the objective includes ensuring high reliability of the power supply by reducing the LPSP, represented by (5.21), and minimizing the excess energy P_{EXC} absorbed by the dummy load P_{Dum}. The P_{EXC} is defined in accordance with (5.24). The objective function is formulated as follows:

$$\text{Min}_X(F) = \text{Min}(\beta_1 EC + \beta_2 LPSP + \beta_3 P_{EXC}) \tag{5.29}$$

In these equations, β implies the weighting coefficient for each objective, while X symbolizes the decision variables associated with the optimization issue that need to be optimized through the applied approaches

$$X = [N_{PV} \ N_{DG} \ N_{batt}] \tag{5.30}$$

5.4   Selected location and load profile

The Coastal Residential Community in Yanbu, located in the Medina Province of Saudi Arabia, located at 24.16° north latitude and 37.32° east longitude, represents an ideal case study for the implementation of a comprehensive renewable energy and desalination system. The selection of this site is based on several logical reasons, primarily centered around its potential for harnessing RESs, specifically PV and wind, as well as the critical need for a sustainable water supply through desalination. The reasons for site selection are Yanbu's abundant solar resources, consistent wind energy potential, acute water desalination needs, substantial energy demand of the residential community, and alignment with environmental sustainability goals. Yanbu, situated in Saudi Arabia's arid region, benefits from significant year-round sunlight, ideal for PV systems, and moderate coastal wind speeds, suitable for wind turbines. The area's proximity to seawater supports essential desalination processes. The high electricity demand of the coastal residential community can be reliably met through these RESs. This approach also supports Saudi Arabia's vision for a sustainable, environmentally friendly future. The renewable energy resources for wind speed and solar radiation in the selected location is shown in Figure 5.3.

The fundamental loads in this area are made up of both household and public utility loads. The simulation system's average load value is 550 kW with a peak value of 1600 kW. Figure 5.4 shows the load demand profile curves on an annual basis. Concerning the data of solar radiation acquired for the investigated locality in Coastal Residential Community in Yanbu, Saudi Arabia, they are available, based on Renewables.ninja [41]. Figure 5.5 shows the annual

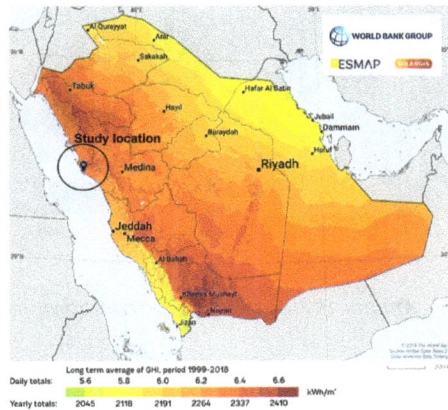

Figure 5.3   Renewable resources in Saudi Arabia: solar horizontal irradiation atlas (source: Global Solar Atlas, 2024 [40])

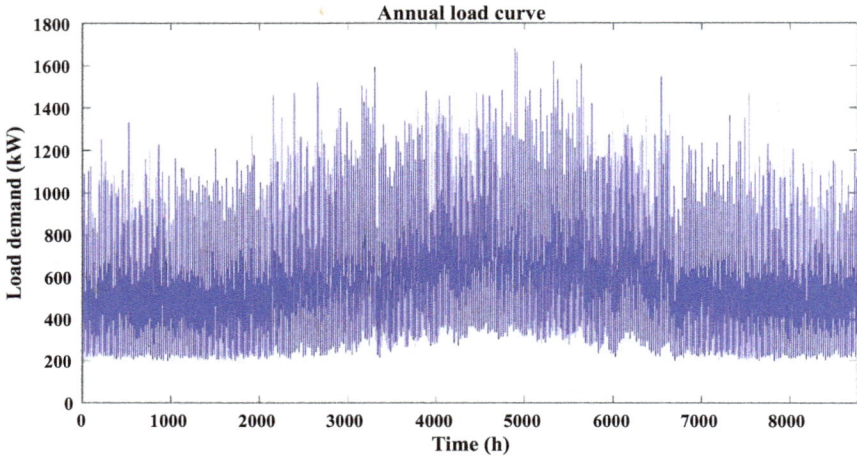

Figure 5.4   Annual profile of load demand for 8760 h

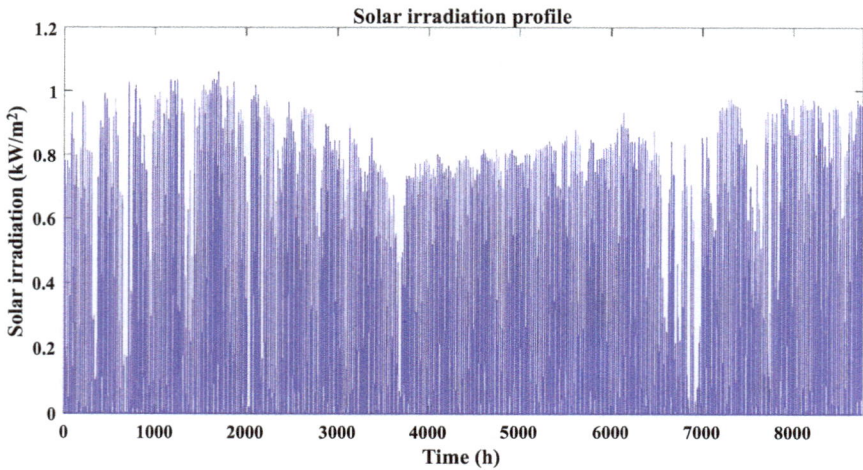

Figure 5.5   Annual variation of solar radiation for 8760 h

solar radiation in the selected site and the annual average radiation is roughly 5.22 kWh/day. As shown in Figure 5.5, the annual average wind speed is roughly 6.19 m/s. The variation of the air temperature in the site is provided in Figure 5.6.

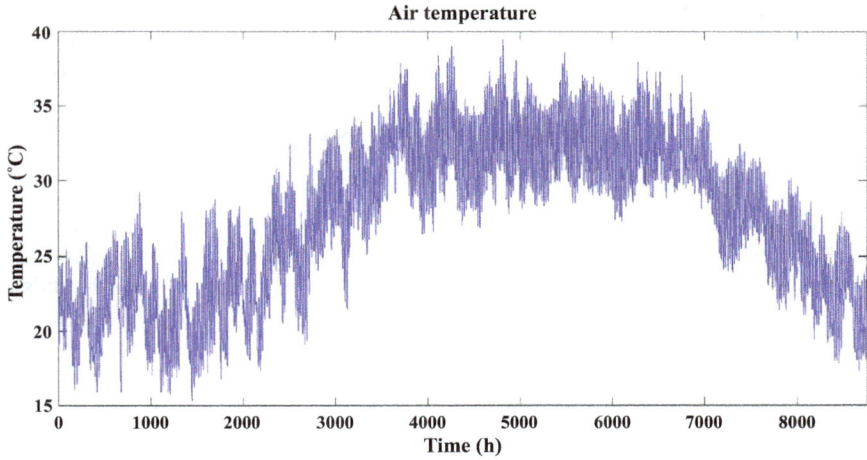

Figure 5.6   Annual variation of air temperature 8760 h

5.5   Optimization algorithms

5.5.1   HO algorithm

The hippopotamus, a semi-aquatic mammal native to Africa, spends much of its life in rivers and ponds, often living in groups called pods or bloats of around 10–30 members. Despite their herbivorous diet of grasses and plants, hippos are highly territorial and can be aggressive, making them one of the most dangerous mammals. Males can weigh around 9920 pounds, while females average up to 3000 pounds, with both genders displaying powerful jaws and a defensive nature against predators. When threatened, they emit a loud warning sound and may flee at velocities of up to 30 km/h to reach water, their primary refuge [24].

5.5.1.1   Inspiration

The HO model takes inspiration from three key behaviors in hippos' lives: social structure, defensive actions, and escape tactics. Hippo groups typically include females, young, multiple males, and a dominant male leader, with curious calves sometimes wandering from the group and risking predation. When threatened, hippos defend themselves by facing predators, using their strong jaws and loud vocalizations, deterring most attackers like lions and hyenas. Finally, if defense fails, hippos instinctively flee to nearby water, where predators are less likely to follow.

5.5.1.2   Mathematical modelling of HO algorithm

The HO algorithm is a population-based approach where each search agent, or "hippopotamus," denotes a candidate solution to the optimization issue. In this algorithm, each hippo's position in the search space corresponds to values of decision variables, represented as vectors, and the collective population forms a matrix. Initial solutions are randomly generated for each variable within its bounds, using a formula involving random numbers between 0 and 1. With a population size of N and m decision variables, the population is mathematically organized into a matrix [24]

$$\chi_i : x_{ij} = lb_{yj} + r.(\text{ubj} - lb_j), i = 1, 2, \ldots, N, j = 1, 2, \ldots, m \qquad (5.31)$$

Here, χ_i represents the location of the $i - th$ candidate solution, r denotes a random number in the range [0 1], and lb and ub are the lower and upper bounds of the $j - th$ decision variable, respectively. N represents the size of hippopotamuses in the group, and m denotes the number of control variables in the optimization issue, forming the population matrix as follows:

$$\chi = \begin{bmatrix} \chi_1 \\ \vdots \\ \chi_i \\ \vdots \\ \chi_N \end{bmatrix}_{N \times m} = \begin{bmatrix} x_{1,1} & \cdots & v_{1,} & \cdots & x_{1,m} \\ \vdots & \ddots & \vdots & \cdot & \vdots \\ x_{i1} & \cdots & x_{y} & \cdots & x_{h,m} \\ \vdots & \cdot & \vdots & \ddots & \vdots \\ x_{N,1} & \cdots & x_N & \cdots & x_{N,m} \end{bmatrix}_{N \times m} \qquad (5.32)$$

Phase 1: Updating hippopotamus positions in the water (exploration). A hippo herd typically includes several adult females, young calves, multiple adult males, and a dominant male who leads the group. This leader is selected based on the best fitness value. Hippos gather closely, with the dominant male protecting the group and territory while adult females position themselves around the males. As males mature, they are often expelled by the dominant male and must either attract mates or compete with other males to assert dominance. Equation (5.33) defines the mathematical update for male hippos' positions in the water

$$X_i^{Mhippo} : x_{ij}^{Mhippo} = x_{ij} + y_1 \cdot \left(D_{hippo} - I_1 x_{ij}\right)$$

$$for \ i = 1, 2, \ldots, \left[\frac{N}{2} \text{ and } j = 1, 2, \ldots, m\right] \qquad (5.33)$$

In (5.33), X_i^{Mhippo} represents the position of a male hippopotamus, while D_{hippo} denotes the position of the dominant hippo—the one with the optimal cost in the current iteration. The term $\vec{r}_{1,\ldots,4}$ implies an arbitrary vector with values ranged from 0 to 1, and r_5 implies a single arbitrary number in the same range (as shown in (5.34)), while I_1 and I_2 are integers between 1 and 2 (Eqs. 5.33 and 5.36). MG_i represents the average position of some randomly chosen hippos, including the current one, x_i, with equal probability, and y_1 is also an arbitrary number with values in the range [0 1] (5.33). In (5.34), e_1 and e_2 indicate random integers that can either be 0 or 1

$$
h = \begin{cases} I_2 \times \vec{r}_1 + (\sim e_1) \\ 2 \times \vec{r}_2 - 1 \\ \vec{r}_3 \\ I_1 \times \vec{r}_4 + (\sim e_2) \\ r_5 \end{cases} \tag{5.34}
$$

$$
T = \exp\left(-\frac{l}{T}\right) \tag{5.35}
$$

$$
X_i^{FBhippo} : x_{ij}^{FBhippo} = \begin{cases} x_{ij} + h_1 \cdot (D_{hippo} - I_2 MG_i) & T > 0.6 \\ [I] & \text{else} \end{cases} \tag{5.36}
$$

$$
[I] = \begin{cases} x_{ij} + h_2 \cdot (I_2 MG_{ij} - D_{hippo}) & r_6 > 0.6 \\ lb_j + r_7 \cdot (ubj - lb_j) & \text{else} \end{cases} \tag{5.37}
$$

$$
\text{for } i = 1, 2, \dots, \left[\frac{N}{2}\right] \text{ and } j = 1, 2, \dots, m
$$

Equations (5.36) and (5.37) outline how the positions of female and immature hippos $X_i^{FBhippo}$ are determined within the herd. Generally, immature hippos stay close to their mothers but may occasionally wander away out of curiosity. When T exceeds 0.6, it suggests that an immature hippo has moved further from its mother (5.35). If r_6, a random number in the range [0 1] (5.37), is higher than 0.5, the immature hippo is still within or near the herd; otherwise, it has moved further away. Equations (5.36) and (5.37) model this behavior, with h_1 and h_2 randomly chosen from five possible scenarios in the h equation, while r_7 in (5.37) denotes another random value ranged from 0 to 1. Equations (5.38) and (5.39) describe position updates for both male and female or immature hippos, with F_i representing the objective function value

$$
X_i = \begin{cases} X_i^{Mhippo}, & F_i^{Mhippo} < F_i \\ X_i & \text{else} \end{cases} \tag{5.38}
$$

$$
X_i = \begin{cases} X_i^{FBhippo}, & F_i^{FBhippo} < F_i \\ X_i & \text{else} \end{cases} \tag{5.39}
$$

The use of h vectors and I_1 and I_2 scenarios strengthen the global search and boost exploration in the optimization approach, ultimately improving its ability to search globally and enhance the exploration process.

Phase 2: Hippopotamus defense against predators (exploration). Herding provides hippos with increased safety, as their large size and numbers often discourage predators from approaching. However, curious young hippos may sometimes wander from the group, making them vulnerable to Nile crocodiles, lions, and hyenas, especially since they lack the strength of adults. Similarly, sick hippos are more susceptible to predators. The primary defense strategy of hippos involves quickly turning toward the threat and emitting loud vocalizations to intimidate it, sometimes even moving closer to drive it away. Equation (5.40) defines the predator's location in the search area [24].

$$Predator : Predator_j = lb_j + \vec{r}_8.(ub_j - lb_j), j = 1, 2, \ldots, m \qquad (5.40)$$

where $\vec{r}_8$ signifies a random vector that varies between [0 1].

$$\vec{D} = |Predator_j - x_{ij}| \qquad (5.41)$$

Equation (5.41) represents the distance from the i-th hippopotamus to a predator. In response to this distance, the hippopotamus exhibits defensive behavior influenced by the factor $Predator_j$ to ensure its safety against potential threats. If $Predator_j$ is less than F_i, it indicates that the predator is very close to the hippopotamus. In this scenario, the hippopotamus reacts quickly by turning towards the predator and advancing towards it, aiming to intimidate the predator into retreating. Conversely, if $Predator_j$ is greater than F_i, this suggests that intruder is at a safer distance from the hippopotamus's field, as illustrated in (5.42). In such a case, the hippopotamus also turns toward the intruder but restricts its movement, signaling its awareness of the predator's presence while maintaining a more cautious stance to protect its territory. This behavior not only reflects the hippopotamus's instinctual response but also serves to establish its dominance and territorial boundaries in the face of potential threats

$$X_i^{HippoR} : x_{ij}^{HippoR} = \begin{cases} RL \oplus predator_j + \left(\dfrac{f}{c - d \times \cos(2\pi g)}\right) \cdot \left(\dfrac{1}{D}\right), & F_{predator_j} < F_i \\ \vec{RL} \oplus predator_j + \left(\dfrac{f}{c - d \times \cos(2\pi g)}\right) \cdot \left(\dfrac{1}{2 \times \vec{D}} + \vec{r}_9\right), & F_{predator_j} \geq F_i \end{cases}$$
$$i = \left[\dfrac{N}{2}\right] + 1, \left[\dfrac{N}{2}\right] + 2, \ldots, N, j = 1, 2, \ldots, m$$

$$(5.42)$$

X_i^{HippoR} represents the position of a hippopotamus that is oriented toward the intruder. $\vec{RL}$ implies an arbitrary vector that follows a Lévy distribution, which is used to account for abrupt changes in the intruder's location at the time of an attack on the hippopotamus. The mathematical expression for Lévy flight is described in (5.43). Here, ω and ϑ are random numbers within [0,1], while ϑ is a constant set at ($\vartheta = 1.5$). The symbol Γ denotes the Gamma function, and the value of σ_ω can be determined using (5.44)

$$Levy(\vartheta) = 0.05 \times \frac{\omega \times \sigma_\omega}{|\vartheta|^{\frac{1}{\vartheta}}} \qquad (5.43)$$

$$\sigma_\omega = \left| \frac{\Gamma(1 + \vartheta) \sin\left(\frac{\pi\vartheta}{2}\right)}{\Gamma\left(\frac{(1+\vartheta)}{2}\right) \vartheta^{\frac{\vartheta-1}{2}}} \right|^{\frac{1}{\vartheta}} \qquad (5.44)$$

In (5.42), f implies a uniformly distributed arbitrary number ranging from [2 4], c implies a random constant within [1 1.5], and D implies a random constant within the range of [2 3]. The variable g signifies a random number within $[-1\ 1]$ Additionally, r_9 implies a random vector dimensioned in $1 \times m$.

In accordance with (5.45), if F_i^{HippoR} exceeds F, it indicates that the hunted hippopotamus will be replaced by another in the herd. Conversely, if F_i^{HippoR} is not greater than F, the predator will flee, allowing the hippopotamus to return to the group. The first and second phases work in tandem to effectively reduce the risk of becoming stacked at the local minimum

$$X_i = \begin{cases} X_i^{HippoR}, & F_i^{HippoR} < F_i \\ X_i, & F_i^{HippoR} \geq F_i \end{cases} \tag{5.45}$$

Phase 3: Hippopotamus evasion from the predator (exploitation). Another reaction of a hippopotamus when faced with a predator occurs when it encounters a group of predators or is unable to successfully fend off an attacker with its defensive tactics. In such cases, the hippopotamus attempts to distance itself from the threat [24]. Typically, the hippopotamus seeks refuge in the nearest lake or pond, as spotted lions and hyenas tend to avoid entering water. This tactic enables the hippopotamus to find a safer spot close to its original location, and modeling this behavior in Phase 3 of the optimization process enhances the capacity for local search exploitation. To model this escape response, a random location is produced in proximity to the hippopotamus's present location. This behavior is described by (5.46)–(5.49). If the newly generated position results in an improved fitness value, it suggests that the hippopotamus has successfully identified a safer location nearby and has adjusted its position accordingly. Here, t implies the present iteration, while T presents the termination criterion, *MaxIter*

$$lb_j^{local} = \frac{lb_j}{t}, ub_j^{local} = \frac{ub_j}{t}, t = 1, 2, \ldots, T \tag{5.46}$$

$$X_i^{Hippo\varepsilon} : x_{ij}^{Hippo\varepsilon} = x_{ij} + r_{10} \cdot \left(lb_j^{local} + s_1 \cdot (ub_j^{local} - lb_j^{local})\right) \tag{5.47}$$
$$i = 1, 2, \ldots, N, j = 1, 2, \ldots, m$$

In (37), $X_i^{Hippo\varepsilon}$ denotes the position of the hippopotamus that is explored to identify the nearest safe location. s_1 is a random vector or number chosen from one of three scenarios, as described in (5.48). These scenarios s facilitate a more effective local search, thereby enhancing the overall exploitation capability of the proposed algorithm

$$s = \begin{cases} 2 \times \vec{r}_{11} - 1 \\ r_{12} \\ r_{13} \end{cases} \tag{5.48}$$

In (5.48), $\vec{r}_{11}$ signifies a random vector with values between [0 1], while r_{10} (as shown in (5.47)) and r_{12} represent random numbers generated within the same range. Furthermore, r_{12} implies an arbitrary number that follows a normal distribution

$$X_i = \begin{cases} X_i^{Hippo\varepsilon}, & F_i^{Hippo\varepsilon} < F_i \\ X_i, & F_i^{Hippo\varepsilon} \geq F_i \end{cases} \qquad (5.49)$$

In the HO algorithm, we opted not to categorize the group into three distinct groups of immature, female, and male hippopotamuses. While such a division would provide a more accurate representation of their behavior, it could hinder the optimization algorithm's performance.

After each iteration of the HO technique, all candidates of the population are updated in accordance with Phases 1–3. This updating mechanism, guided by (5.33)–(5.49), continues until the program termination. Throughout the algorithm's implementation, the best possible solution is continuously monitored and recorded. Once the algorithm has concluded, the optimal solution, referred to as the dominant hippopotamus solution, is revealed as the final outcome. The procedural details of the HO algorithm are illustrated in the flowchart shown in Figure 5.7.

5.5.2 *GGO algorithm*

The GGO approach draws inspiration from the social behaviors and dynamic activities of geese. Geese form strong, lifelong bonds with their mates and are protective of their young. They often remain close to an ill or injured partner, even as the rest of the flock migrates to hotter areas for the winter. If a goose loses its partner, it may isolate itself and remain single for life. Geese display interesting behaviors while foraging for food and building nests, gathering materials like leaves and sticks [25].

During the spring, male geese guard their nests while females incubate the eggs for about 30 days. Many geese return to the same nesting sites each year. Families come together to form a gaggle, where members look out for each other, with one or two designated sentries watching for predators while others eat. These roles rotate, similar to sailors on lookout duty. Healthy geese often care for injured flockmates, creating a support system for finding food and evading threats.

Ducks, on the other hand, are social creatures that thrive in groups called "paddlings" when on water. They forage in grassy areas and shallow waters by day and rest together at night. Geese are also remarkable fliers, capable of migrating thousands of kilometers in large flocks, often forming a "V" shape to reduce air resistance, enabling them to travel significantly farther as a group than individuals (Figure 5.8). In case when those at the front become tired, they rotate to the back, allowing others to lead. Their excellent memory helps them navigate during migrations using familiar landmarks and stars.

The GGO methodology begins by randomly generating multiple individuals. Each individual represents a potential solution to the issue at hand. The GGO population is presented as $X_i(i = 1, 2, ..., n)$, where n signifies the size of the gaggle. A fitness function, F_n, is selected to evaluate the solutions within the

Start

Set parameters of hippopotamus number and Maxiter

Generate initial population

Evaluate fitness value

Update hippopotamus based on the objective function

$i > N/2$?   yes

no

Phase 1   calculate X_i^{Mhippo} and $X_i^{FBhippo}$ by (5.33, 5.36)

$i = i+1$ Update X_i using (5.38, 5.39)

Randomly generate predator position using (5.40)

Phase 2   calculate X_i^{hippoR} according to (5.42)

Update X_i using (5.45)

yes $i < N$?

$i = i+1$

no

Set $i = 1$

Phase 3   calculate X_i^{hippoE} according to (5.47)

Update X_i using (5.49)

yes $i < N$?

$i = i+1$

no

Save the best solution so far

no $t < T$?   yes $t = t+1$ $i = 1$

Output best solution

End

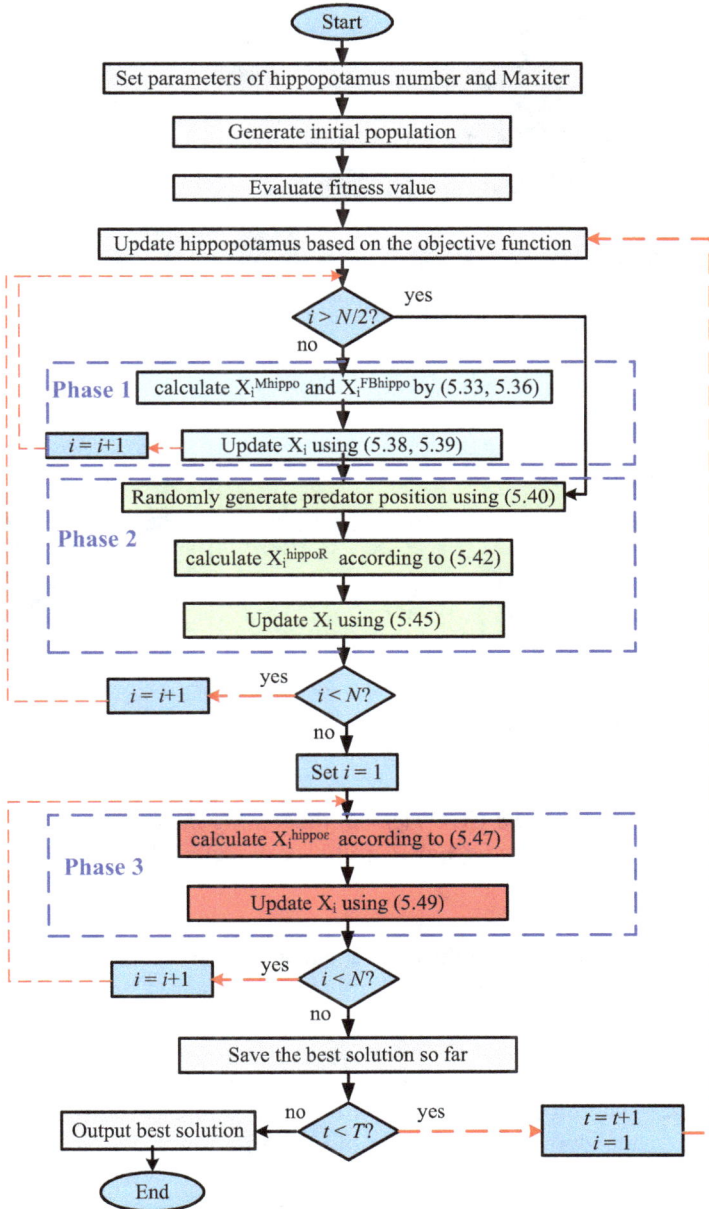

Figure 5.7   *Flowchart of the HO algorithm*

Figure 5.8   Greylag goose approaching its prey (optimal solution) while simultaneously exploring the surrounding area for additional resources [25]

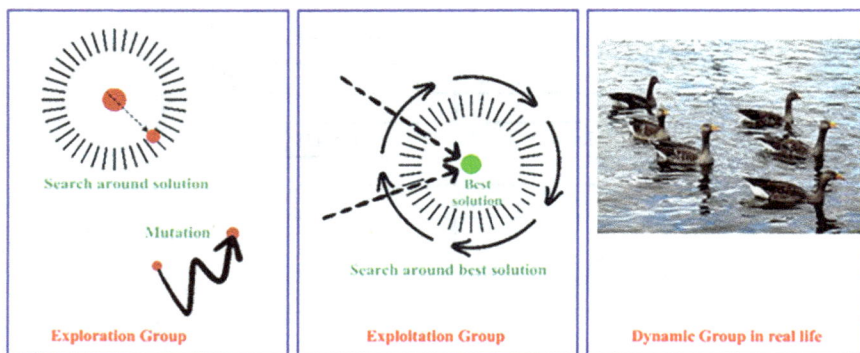

Figure 5.9   Overview of the GGO method, illustrating the exploration, exploitation, and dynamic grouping processes [25]

group. After assessing each individual agent X_i using the fitness function, the optimal solution, referred to as the leader, is identified and labeled as P.

The GGO approach employs a dynamic groups mechanism, dividing candidates into two categories: exploration group n_1 and exploitation group n_2. The number of solutions in each group is dynamically adjusted in each iteration based on the best solution found. Figure 5.9 illustrates the processes of the exploration group with n_1 agents and the exploitation group with n_2 agents. Initially, GGO allocates 50% of the agents to exploration and 50% to exploitation. As iterations proceed, the count of agents in the exploration group n_1 is reduced while the count in the exploitation group n_2 increases. However, if there is no improvement in the optimal solution's fitness value for three consecutive iterations, the approach increases the number of candidates in the exploration group n_1 to seek new optimal solutions and eliminate the risk of getting stacked in local optima.

5.5.2.1 Exploration phase

As elaborated further below, exploration plays a crucial role in identifying promising areas within the problem field and in preventing stagnation at local optima by directing movement toward the optimal solution.

Approaching the best solution: This strategy allows the geese explorers to investigate potentially favorable new locations in their vicinity. They accomplish this by continuously evaluating various nearby options to identify the one with the highest fitness value. The GGO approach employs the upcoming equations to update the A and C vectors as $A = 2a \cdot r_1 - a$ and $C = 2 \cdot r_2$ with advance in the optimization process, with the parameter a decreasing from 2 to 0:

$$X(t+1) = X^*(t) - A \cdot |C \cdot X^*(t) - X(t)| \tag{5.50}$$

Here, $X(t)$ denotes an agent at iteration t, while $X^*(t)$ indicates the location of the optimal solution (leader). The updated location of the agent is given by $X(t+1)$. The values of r_1 and r_2 are randomly chosen within the range $[0, 1]$.

The subsequent equation uses three randomly chosen search agents (paddlings), labeled $X_{Paddle1}$, $X_{Paddle2}$, and $X_{Paddle3}$, to enhance exploration by preventing agents from relying solely on a single leader's position. The present search agent's position is modified as follows for $|A| \geq 1$ [25]:

$$X(t+1) = \omega_1 \times X_{Paddle1} + z \times \omega_2 \times (X_{Paddle2} - X_{Paddle3}) + (1 - z) \times \omega_3$$
$$\times (X - X_{Paddle1}) \tag{5.51}$$

In this equation, the weights ω_1, ω_2, and ω_3 are updated within the range $[0, 2]$. The z decreases exponentially and is obtained as follows:

$$z = 1 - \left(\frac{t}{t_{max}}\right)^2 \tag{5.52}$$

where t signifies the current iteration and t_{max} represents the iterations.

In the second updating mechanism, where the values of a and A are decreased for $r_3 \geq 0.5$, the update is given by

$$X(t+1) = \omega_4 \times |X^*(t) - X(t)| \cdot e^{bl} \cdot \cos(2\pi l) + [2\omega_1(r_4 + r_5) \times X^*(t)] \tag{5.53}$$

Here, b implies a constant, and l implies a random constant within the range $[-1, 1]$. The weight parameter ω_4 is updated within $[0, 2]$, while r_4 and r_5 are updated within $[0, 1]$.

5.5.2.2 Exploitation phase

The exploitation team focuses on refining existing solutions. By the end iteration, the GGO identifies the individual with the highest fitness and rewards them.

To fulfil its exploitation goals, the GGO employs two distinct approaches outlined below.

Approaching the best solution: The GGO utilizes the following equations to guide the movement toward the best solution. Three solutions, referred to as sentries—$X_{Sentry1}$, $X_{Sentry2}$, and $X_{Sentry3}$—assist other individuals $X_{NonSentry}$ in adjusting their positions toward the estimated location of the optimal solution. The position updates are computed as follows:

$$
\begin{aligned}
X_1 &= X_{Sentry1} - A_1 \cdot \left| C_1 \cdot X_{Sentry1} - X \right| \\
X_2 &= X_{Sentry2} - A_2 \cdot \left| C_2 \cdot X_{Sentry2} - X \right| \\
X_3 &= X_{Sentry3} - A_3 \cdot \left| C_3 \cdot X_{Sentry3} - X \right|
\end{aligned}
\tag{5.54}
$$

where A_1, A_2, and A_3 are calculated as $A = 2a \cdot r_1 - a$ and C_1, C_2, C_3 as $C = 2 \cdot r_2$. The modified locations for the population $X(t+1)$ can be defined as the mean of the three solutions:

$$
X(t+1) = \bar{X}_i \big|_0^3
\tag{5.55}
$$

Exploring around the best solution: Individuals are encouraged to search for enhancements in areas close to the leader (best solution). This leads some agents to investigate regions near the optimal response, denoted as X_{Flock1}. The GGO implements this process with the following formula:

$$
X(t+1) = X(t) + \mathrm{D}(1+z) \times \omega \times (X - X_{Flock1})
\tag{5.56}
$$

5.5.2.3   Selection of the optimal solution

The GGO employs a mutation technique and examines members of the exploration group to enhance its exploration capabilities. This strong ability to explore allows the GGO to delay convergence.

The process begins by inputting data into the GGO, including parameters such as population size, mutation rate, and the number of iterations. The GGO subsequently separates participants into two groups: those engaged in exploration and those focused on exploitation. During the iterative process of finding the optimal solution, the GGO dynamically modifies the sizes of these groups. Each group utilizes two strategies to fulfill its objectives. To ensure diversity and thorough investigation, the GGO randomly reshuffles the solutions between iterations. For instance, a solution from the exploration group may transition to the exploitation group in the next iteration.

The elitist strategy of the GGO ensures that the leader remains in place throughout the process. As illustrated in Figure 5.9, the steps of the GGO algorithm are executed to update the positions of the exploration group n_1 and the exploitation group n_2. The parameter r_1 is updated during iterations according to the formula $r_1 = c\left(1 - \frac{t}{t_{max}}\right)$, where t denotes the present iteration, c implies a constant, and t_{max} denotes the max number of iterations. By the end of every iteration, the GGO modifies the agents in the search field and randomly changes their order to swap roles between the exploration and exploitation groups. Finally, the GGO outputs the best solution identified.

5.6 Results and discussion

The optimization programs utilizing the HO and GGO algorithms were executed using Matlab software. The program ran for 200 iterations with 30 search agents. Figure 5.10 illustrates the convergence curves from the optimization process aimed at achieving the optimal objective function (as defined in (30)) for the proposed hybrid energy system, which includes PV, DG, and battery storage. The results indicate that the HO algorithm outperforms the GGO algorithm in terms of fitness value and convergence behavior. The results of the optimization process based on both HO and GGO algorithms are tabulated in Table 5.2. The results from the

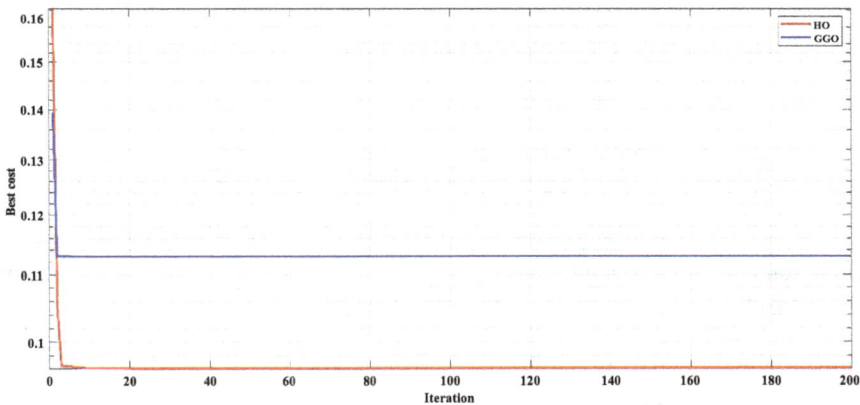

Figure 5.10 Convergence characteristics of both HO and GGO algorithms

Table 5.2 Results of the optimal design process based on HO and GGO algorithms

		HO	GGO
Best fitness		0.096265297	0.1129050920
Best solution	N_{PV} (unit)	1966	2478
	N_{DG} (unit)	8	7
	N_{Batt} (unit)	4534	8448
Performance characteristics	COE ($/kWh)	0.329722877	0.4210802889
	NPC ($)	21,307,268.35	27,210,943.916
	LPSP	0.036731823	0.0221884971
	EXP	0.039849367	0.0586492856
	Fuel consumption (liter/year)	997,784.5213	805,908.49787
	Fuel cost ($/year)	798,227.6171	644,726.79829
	Annual cost ($/year)	1,666,797.677	2,128,622.8420

optimal design process utilizing the HO and GGO algorithms demonstrate that the HO method surpasses GGO in several important metrics. The best fitness value for HO is 0.096265297, while GGO achieves 0.1129050920. This comparison highlights the superior performance of the HO method in this optimization scenario. Furthermore, HO offers a more favorable net present value (NPV) of 1966 compared to GGO's 2478, as well as a lower COE at $0.329722877 versus GGO's $0.4210802889. Although the HO method necessitates a greater number of distributed generators (N_{DG}) and fewer batteries (N_{Batt}), it results in reduced total annual costs and fuel expenses, underscoring its enhanced efficiency and economic feasibility in designing hybrid energy systems. Based on the results obtained from the HO algorithm, the proposed hybrid system is operated for one year and the performance of the system is evaluated. Figure 5.11 presents the flow of electrical power between the system components. Figure 5.12 shows the variation of the SOC of the battery storage system over the year, while Figure 5.13 presents the hourly fuel consumption of the DG over one week of operation.

Figure 5.14 depicts the fluctuation of LPS concerning load demand throughout the year. The red line represents instances of LPS, emphasizing times when the power generated does not meet demand. The numerous spikes indicate moments of supply shortfalls. A distinct correlation is evident between load demand and LPS; as demand peaks, especially during periods of high consumption, LPS rises as well, implying that the capacity for power generation may not be sufficient to satisfy the load requirements.

Figure 5.15 depicts power dissipation in a dummy load during times of excess generation compared to load demand. When power generation surpasses the actual load needs, the surplus energy is redirected to the dummy load, which helps prevent system overload and maintain stability. The power absorbed by the dummy load exhibits intermittent spikes, reflecting instances when excess energy is taken in. This primarily happens when load demand is elevated but still exceeded by

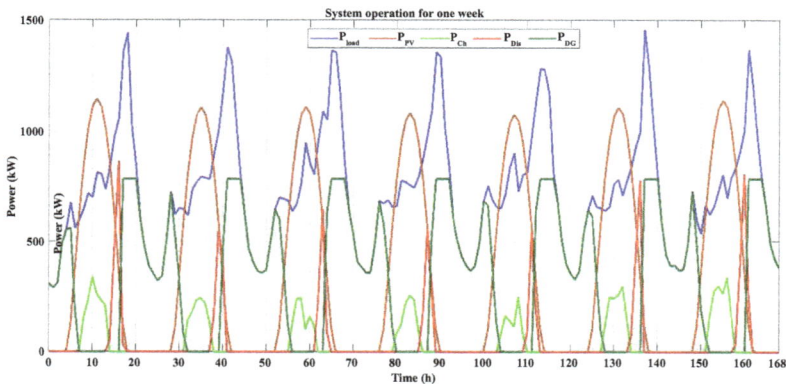

Figure 5.11   System performance over one week of operation

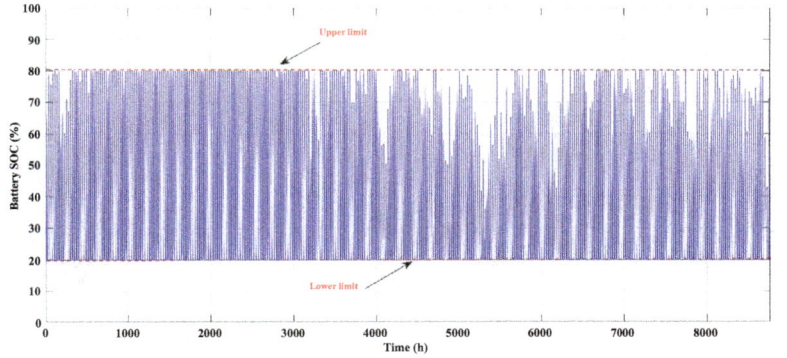

(a)

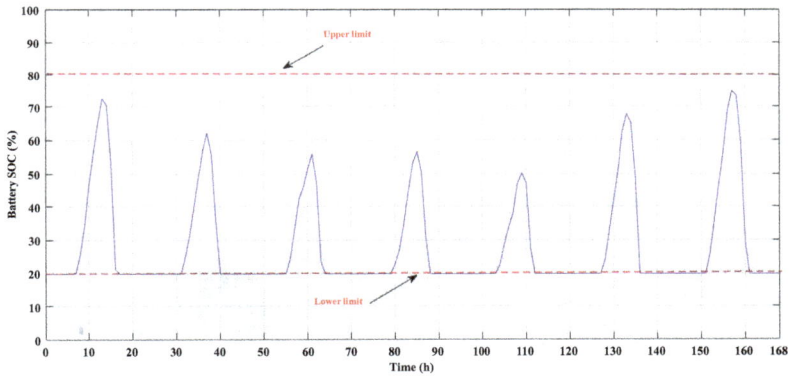

(b)

Figure 5.12   SOC of the battery bank: (a) over the year and (b) for one week

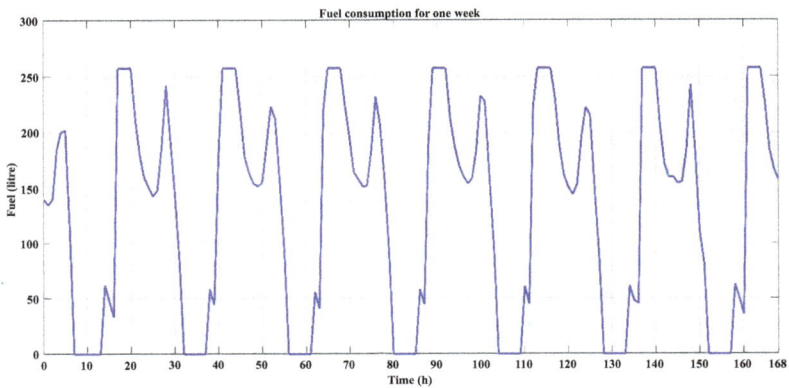

Figure 5.13   Variation of DG fuel consumption per hour over one week

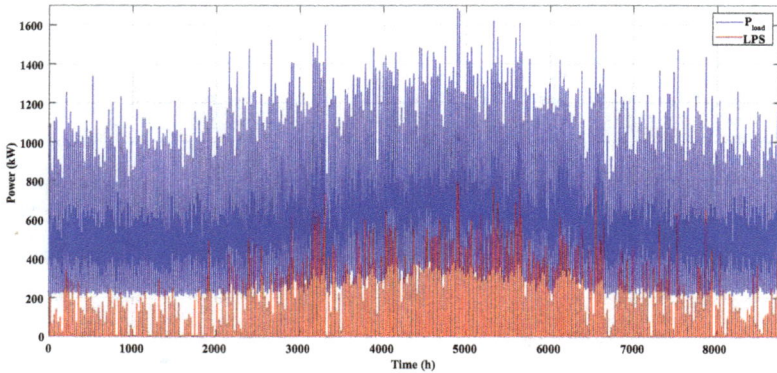

Figure 5.14   Variation of LPS with respect to load demand over the year

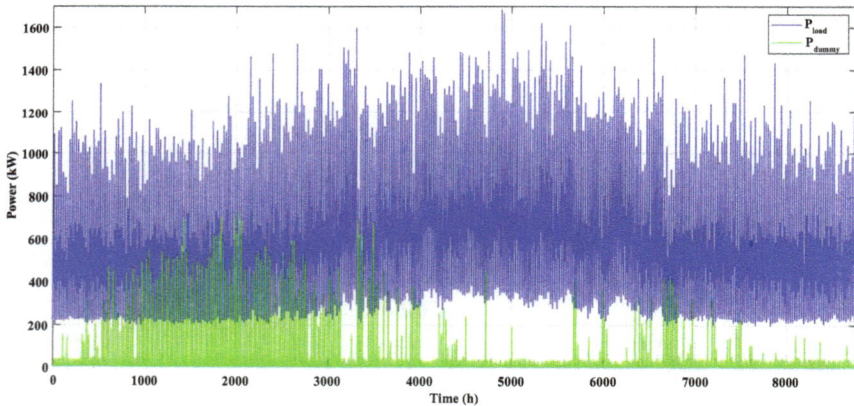

Figure 5.15   Variation of power consumed in dumper load with respect to load demand over the year

generation, leading to overflow directed to the dummy load. This strategy enhances system efficiency by effectively managing surplus energy.

Based on the results obtained from the HO and GGO algorithms, the division of the annual cash flow in the system is tabulated in Table 5.3. It can be noticed that the GGO algorithm incurs higher costs for the PV system, battery, and inverter than the HO algorithm, likely due to expanded solar capacity and enhanced storage and integration components. This shift points to a greater reliance on renewable energy, with a corresponding reduction in DG costs, lowering fuel consumption and environmental impact. While the GGO setup has a total annual cost of $2.13 million, higher than the $1.67 million with HO, it reflects a strategic trade-off, prioritizing sustainability by investing more in renewables and storage to decrease diesel dependence. In contrast, the HO configuration emphasizes a balanced approach, relying more on diesel, leading to lower initial investments but higher ongoing fuel expenses.

The division of the annual cost of each subsystem is provided in Table 5.4. The results are also presented graphically in Figure 5.16 regarding the type of generating technology utilized in each subsystem, while Figure 5.17 shows the

Table 5.3 Distribution of the annual cost of the system between the system components

	PV system ($/year)	DG ($/year)	Battery ($/year)	Inverter ($/year)	Total cost ($/year)
HO	87,331.05006	881,782.5658	258,094.96240	439,589.09952	1,666,797.677
GGO	110,080.01874	718,469.11350	480,937.56695	819,136.14282	2,128,622.842

Table 5.4 Division of the annual cost based on the type of cost in each subsystem based on the HO algorithm

	Capital	O&M	Fuel	Replacement	Total
PV system	77,502.28681$	9828.763248$	0	0	87,331.05$
Battery banks	191,676.8942$	19,925.42045$	0	46,492.64773$	258,095$
DG	42,699.54287$	9428.170578$	798,227.61711$	31,427.23526$	881,782.6$
Inverter	262,776.4947$	0	0	176,812.6047$	439,589.1$
Total annual cost					1,666,798$

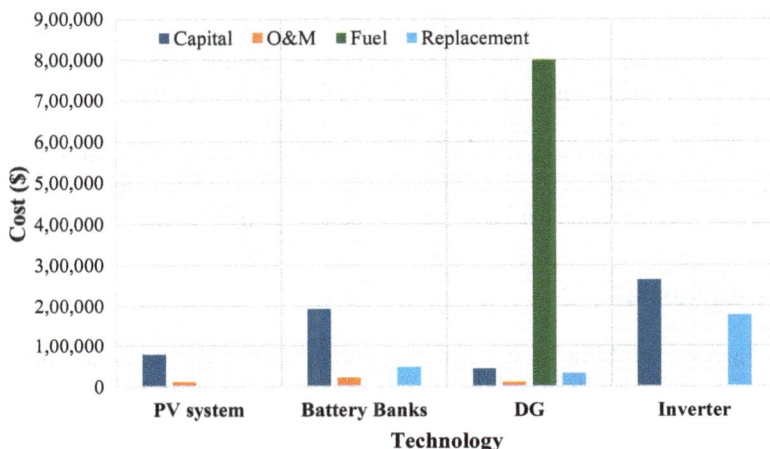

Figure 5.16 Distribution of the annual cost of the system based on used technology

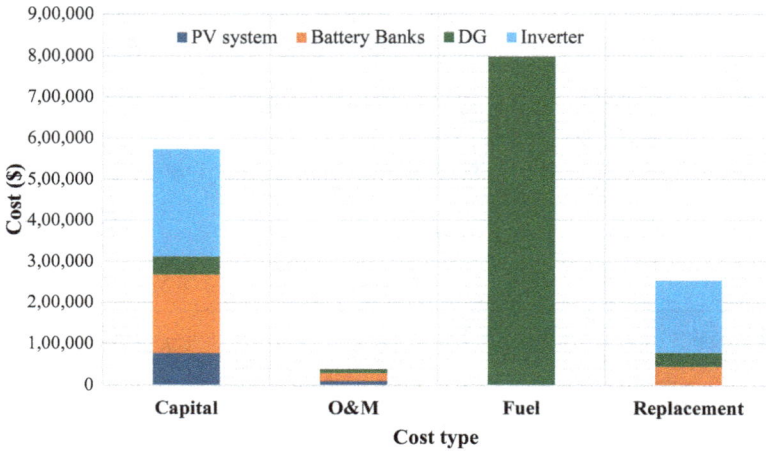

Figure 5.17   Distribution of the annual cost of the system based on cost type

distribution of cash flow with respect to the cost type. The inverter and battery banks demand the highest capital outlay, reflecting the substantial expenses tied to energy storage and conversion. In contrast, the PV system has a lower upfront cost, consistent with typically lower equipment expenses. Battery banks also incur the most significant O&M costs due to regular maintenance needs for energy storage, while both the DG and PV system show moderate O&M expenses. The inverter, on the other hand, has no O&M costs, possibly due to a maintenance-free design or unaccounted maintenance. Fuel costs for the DG are considerable, making it the costliest subsystem overall. This reliance on diesel fuel significantly raises annual expenses, highlighting the financial and environmental burden of fossil fuel dependence in the hybrid system. The PV and battery components, however, naturally avoid fuel costs. Replacement costs for the inverter and batteries are also notable, primarily due to component wear and degradation. Batteries, with shorter lifespans, especially contribute to frequent replacements and associated costs, whereas PV panels generally do not require replacement over the analyzed period. The DG's high fuel use drives its total cost to over half of the system's annual expenditure, with inverter and battery banks further adding to expenses through high initial and replacement costs. The PV system remains the most economical, reinforcing its value as a cost-effective component in hybrid systems.

5.7   Conclusion

In this chapter, we have implemented the optimal configuration and component sizing of a proposed hybrid energy system over a 25-year timeframe. This hybrid system incorporates solar PV panels, a DG, a battery storage system, and an inverter to facilitate the efficient transfer of generated electrical energy to users. We addressed the complex optimization challenges associated with this hybrid setup, employing both the HO and the GGO. The optimization process successfully identified the best parameters for the subsystems that constitute the overall system. Notably, the proposed HO algorithm proved to be more effective than the GGO algorithm in solving the optimization problem, demonstrating its robustness in determining optimal values for the various components of the system. From the obtained results, the HO approach resulted in an NPC of 21.3 M$, a 21.6% reduction compared to GGO's 27.2 M$. Additionally, HO yielded a lower COE at 0.33 $/kWh, demonstrating enhanced economic feasibility over GGO's 0.42 $/kWh. In terms of energy reliability, GGO achieved a lower LPSP of 0.022, compared to HO's 0.0367, showing GGO's advantage in this respect. However, the HO configuration reduced excess power directed to the dummy load to 0.0398, compared to GGO's 0.0586. These results illustrate that the HO approach offers a sustainable, efficient strategy for hybrid energy systems, balancing economic and environmental considerations for off-grid rural electrification. Future research could explore further algorithmic refinements and the integration of additional renewable sources to enhance system resilience and efficiency. Furthermore, to enhance the effectiveness of hybrid energy systems, future research could focus on integrating advanced energy management systems and exploring additional RESs. Additionally, conducting life cycle cost analyses and implementing pilot projects will provide valuable insights into the long-term performance and viability of these systems.

Acknowledgment

The authors would like to acknowledge the research support provided by The Renewable Energy Technical Incubator (RETI) funded by the National Industrial Development and Logistics Program (NIDLP) under the Interdisciplinary Research Center for Sustainable Energy Systems (IRC-SES) at King Fahd University of Petroleum and Minerals (KFUPM), through Project No. CREP2522.

References

[1] M. Das, M. A. K. Singh, and A. Biswas, "Techno-economic optimization of an off-grid hybrid renewable energy system using metaheuristic optimization approaches–case of a radio transmitter station in India," *Energy Conversion and Management*, vol. 185, pp. 339–352, 2019.

[2] H. M. Sultan, A. A. Z. Diab, O. N. Kuznetsov, Z. M. Ali, and O. Abdalla, "Evaluation of the impact of high penetration levels of PV power plants on the capacity, frequency and voltage stability of Egypt's unified grid," *Energies*, vol. 12, no. 3, p. 552, 2019.

[3] S. Rehman, K. M. Kotb, M. E. Zayed *et al.*, "Techno-economic evaluation and improved sizing optimization of green hydrogen production and storage under higher wind penetration in Aqaba Gulf," *Journal of Energy Storage*, vol. 99, p. 113368, 2024.

[4] S. Rehman, M. E. Zayed, K. Irshad *et al.*, "Design, commissioning and operation of a large-scale solar linear Fresnel system integrated with evacuated compound receiver: Field testing, thermodynamic analysis, and enhanced machine learning-based optimization," *Solar Energy*, vol. 278, p. 112785, 2024.

[5] S. Rehman, K. Irshad, M. A. Mohandes, A. Al-Shaikhi, and M. E. Zayed, "Comprehensive evaluation of solar floating photovoltaic prospective in Saudi Arabia: Comparative experimental investigation and thermal performance analysis," *Solar Energy*, vol. 283, p. 113015, 2024.

[6] B. Shboul, M. E. Zayed, N. Al-Tawalbeh *et al.*, "Dynamic numerical modeling and performance optimization of solar and wind assisted combined heat and power system coupled with battery storage and sophisticated control framework," *Results in Engineering*, vol. 22, p. 102198, 2024.

[7] M. E. Zayed, J. Zhao, A. H. Elsheikh, *et al.*, "Applications of cascaded phase change materials in solar water collector storage tanks: A review," *Solar Energy Materials and Solar Cells*, vol. 199, pp. 24–49, 2019.

[8] M. E. Zayed, A. E. Kabeel, B. Shboul *et al.*, "Performance augmentation and machine learning-based modeling of wavy corrugated solar air collector embedded with thermal energy storage: Support vector machine combined with Monte Carlo simulation," *Journal of Energy Storage*, vol. 74, p. 109533, 2023.

[9] I. R. E. Agency, "Renewable Energy Statistics 2022," *IRENA*, vol. https://www.irena.org/, 2022.

[10] A. S. Menesy, S. Almomin, H. M. Sultan *et al.*, "Techno-economic optimization framework of renewable hybrid photovoltaic/wind turbine/fuel cell energy system using artificial rabbits algorithm," *IET Renewable Power Generation*, vol. 18, no. 15, pp. 2907–2924, 2024.

[11] A. S. Menesy, H. M. Sultan, I. O. Habiballah, H. Masrur, K. R. Khan, and M. Khalid, "Optimal configuration of a hybrid photovoltaic/wind turbine/biomass/hydro-pumped storage-based energy system using a heap-based optimization algorithm," *Energies*, vol. 16, no. 9, p. 3648, 2023.

[12] H. M. Sultan, A. S. Menesy, S. Kamel, A. S. Alghamdi, and M. Zohdy, "Optimal sizing of isolated hybrid PV/WT/FC system using manta ray foraging optimization algorithm," *International Transaction Journal of Engineering, Management, & Applied Sciences & Technologies*, vol. 11, pp. 1–12, 2020.

[13] H. M. Sultan, A. S. Menesy, S. Kamel, A. S. Alghamdi, and C. Rahmann, "Optimal design of a grid-connected hybrid photovoltaic/wind/fuel cell system," in *2020 IEEE Electric Power and Energy Conference (EPEC)*, 2020, pp. 1–6: IEEE.

[14] H. M. Sultan, A. S. Menesy, S. Kamel, A. Korashy, S. Almohaimeed, and M. Abdel-Akher, "An improved artificial ecosystem optimization algorithm for optimal configuration of a hybrid PV/WT/FC energy system," *Alexandria Engineering Journal*, vol. 60, no. 1, pp. 1001–1025, 2021.

[15] M. Thirunavukkarasu, H. Lala, and Y. Sawle, "Reliability index based optimal sizing and statistical performance analysis of stand-alone hybrid renewable energy system using metaheuristic algorithms," *Alexandria Engineering Journal*, vol. 74, pp. 387–413, 2023.

[16]   A. M. Eltamaly, E. Ali, M. Bumazza, S. Mulyono, and M. Yasin, "Optimal design of hybrid renewable energy system for a reverse osmosis desalination system in Arar, Saudi Arabia," *Arabian Journal for Science and Engineering*, vol. 46, pp. 9879–9897, 2021.

[17]   I. B. Ali, M. Turki, J. Belhadj, and X. Roboam, "Systemic design and energy management of a standalone battery-less PV/Wind driven brackish water reverse osmosis desalination system," *Sustainable Energy Technologies and Assessments*, vol. 42, p. 100884, 2020.

[18]   C. Li, Y. Zheng, Z. Li *et al.*, "Techno-economic and environmental evaluation of grid-connected and off-grid hybrid intermittent power generation systems: A case study of a mild humid subtropical climate zone in China," *Energy*, vol. 230, p. 120728, 2021.

[19]   A. S. Aziz, M. F. N. Tajuddin, M. K. Hussain *et al.*, "A new optimization strategy for wind/diesel/battery hybrid energy system," *Energy*, vol. 239, p. 122458, 2022.

[20]   T. A. Ajiwiguna, G.-R. Lee, B.-J. Lim, S.-H. Cho, and C.-D. Park, "Optimization of battery-less PV-RO system with seasonal water storage tank," *Desalination*, vol. 503, p. 114934, 2021.

[21]   W. Khiari, M. Turki, and J. Belhadj, "Power control strategy for PV/Wind reverse osmosis desalination without battery," *Control Engineering Practice*, vol. 89, pp. 169–179, 2019.

[22]   K. Elmaadawy, K. M. Kotb, M.R. Elkadeem *et al.*, "Optimal sizing and techno-enviro-economic feasibility assessment of large-scale reverse osmosis desalination powered with hybrid renewable energy sources," *Energy Conversion and Management*, vol. 224, p. 113377, 2020.

[23]   H. Mehrjerdi, "Modeling and optimization of an island water-energy nexus powered by a hybrid solar-wind renewable system," *Energy*, vol. 197, p. 117217, 2020.

[24]   M. H. Amiri, N. Mehrabi Hashjin, M. Montazeri, S. Mirjalili, and N. Khodadadi, "Hippopotamus optimization algorithm: a novel nature-inspired optimization algorithm," *Scientific Reports*, vol. 14, no. 1, p. 5032, 2024.

[25]   E.-S. M. El-Kenawy, N. Khodadadi, S. Mirjalili, A. A. Abdelhamid, M. M. Eid, and A. Ibrahim, "Greylag goose optimization: nature-inspired optimization algorithm," *Expert Systems with Applications*, vol. 238, p. 122147, 2024.

[26]   W. Dong, Y. Li, and J. Xiang, "Optimal sizing of a stand-alone hybrid power system based on battery/hydrogen with an improved ant colony optimization," *Energies*, vol. 9, no. 10, p. 785, 2016.

[27]   A. A. Z. Diab, H. M. Sultan, T. D. Do, O. M. Kamel, and M. A. Mossa, "Coyote optimization algorithm for parameters estimation of various models of solar cells and PV modules," *IEEE Access*, vol. 8, pp. 111102–111140, 2020.

[28]   A. A. Z. Diab, H. M. Sultan, R. Aljendy, A. S. Al-Sumaiti, M. Shoyama, and Z. M. Ali, "Tree growth based optimization algorithm for parameter extraction of different models of photovoltaic cells and modules," *IEEE Access*, vol. 8, pp. 119668–119687, 2020.

[29] O. M. Kamel, A. A. Z. Diab, M. M. Mahmoud, A. S. Al-Sumaiti, and H. M. Sultan, "Performance enhancement of an islanded microgrid with the support of electrical vehicle and STATCOM systems," *Energies*, vol. 16, no. 4, p. 1577, 2023.

[30] H. Abd El-Sattar, H. M. Sultan, S. Kamel, T. Khurshaid, and C. Rahmann, "Optimal design of stand-alone hybrid PV/wind/biomass/battery energy storage system in Abu-Monqar, Egypt," *Journal of Energy Storage*, vol. 44, p. 103336, 2021.

[31] M. Thirunavukkarasu, Y. Sawle, and H. Lala, "A comprehensive review on opti-mization of hybrid renewable energy systems using various optimization techni-ques," *Renewable and Sustainable Energy Reviews*, vol. 176, p. 113192, 2023.

[32] C. A. W. Ngouleu, Y. W. Koholé, F. C. V. Fohagui, and G. Tchuen, "Techno-economic analysis and optimal sizing of a battery-based and hydrogen-based standalone photovoltaic/wind hybrid system for rural elec-trification in Cameroon based on meta-heuristic techniques," *Energy Con-version and Management*, vol. 280, p. 116794, 2023.

[33] O. Ekren and B. Y. Ekren, "Size optimization of a PV/wind hybrid energy conversion system with battery storage using simulated annealing," *Applied Energy*, vol. 87, no. 2, pp. 592–598, 2010.

[34] I. Amoussou, E. Tanyi, L. Fatma *et al.*, "The optimal design of a hybrid solar PV/Wind/Hydrogen/Lithium battery for the replacement of a heavy fuel oil thermal power plant," *Sustainability*, vol. 15, no. 15, p. 11510, 2023.

[35] A. A. Z. Diab, H. M. Sultan, I. S. Mohamed, O. N. Kuznetsov, and T. D. Do, "Application of different optimization algorithms for optimal sizing of PV/wind/diesel/battery storage stand-alone hybrid microgrid," *IEEE Access*, vol. 7, pp. 119223–119245, 2019.

[36] M. Kharrich, O. H. Mohammed, S. Kamel *et al.*, "Development and imple-mentation of a novel optimization algorithm for reliable and economic grid-independent hybrid power system," *Applied Sciences*, vol. 10, no. 18, p. 6604, 2020.

[37] A. M. Jasim, B. H. Jasim, F.-C. Baiceanu, and B.-C. Neagu, "Optimized sizing of energy management system for off-grid hybrid solar/wind/battery/biogasifier/diesel microgrid system," *Mathematics*, vol. 11, no. 5, p. 1248, 2023.

[38] M. A. Mohamed, A. M. Eltamaly, A. I. Alolah, and A. Hatata, "A novel framework-based cuckoo search algorithm for sizing and optimization of grid-independent hybrid renewable energy systems," *International Journal of Green Energy*, vol. 16, no. 1, pp. 86–100, 2019.

[39] R. Khezri, P. Razmi, A. Mahmoudi, A. Bidram, and M. H. Khooban, "Machine learning-based sizing of a renewable-battery system for grid-con-nected homes with fast-charging electric vehicle," *IEEE Transactions on Sustainable Energy*, vol. 14, no. 2, pp. 837–848, 2022.

[40] G. Atlas, "Global Solar Atlas," https://globalsolaratlas.info/, 2024.

[41] S. Pfenninger and I. Staffell, "Renewables.ninja." 2018. [Online]. Available: https://www.renewables.ninja/ [Accessed 1/9/2024].

Chapter 6

A novel smart grid concept for a 100% green hybrid energy system

Ahmed A. Zaki Diab[1,2]

Abstract

As global efforts to combat climate change and reduce greenhouse gas emissions intensify, renewable energy systems have become a cornerstone of modern energy strategies and environmental policies. This study investigates the potential and challenges of implementing fully green hybrid energy systems, with a particular focus on energy storage as a critical component. Two configurations are analyzed to enhance system performance, reliability, and cost efficiency in a 100% renewable energy microgrid: (1) a photovoltaic (PV) and battery storage system (BSS) and (2) a PV/wind turbine (WT)/BSS system. The research targets the optimization of a hybrid energy system for an industrial community in New Minya, Egypt, utilizing site-specific solar radiation and wind speed data for precise design and simulation. Key economic metrics, including net present cost (NPC), levelized cost of energy (LCOE), and installation, operation, and maintenance costs, are evaluated for each configuration. Simulations and optimizations are conducted using HOMER software, developed by the National Renewable Energy Laboratory. Results indicate that the PV/BSS configuration is both economically viable and environmentally sustainable, with an NPC of $88,429, an LCOE of $0.116 per kWh, and zero carbon emissions. This configuration emerges as a competitive alternative to diesel generator systems, offering significant economic and environmental benefits. Additionally, the PV/WT/BSS system achieved an LCOE of $0.119 per kWh, demonstrating cost-effectiveness while maintaining zero emissions. Both configurations outperform traditional diesel-based systems, which are characterized by higher operational costs and considerable environmental drawbacks.

Keywords: 100% Green hybrid energy system; Renewable energy; Optimization; HOMER

[1]Electrical Engineering Department, Minia University, Egypt
[2]Department of Mechatronics Engineering, Minia National University, Egypt

6.1   Introduction

Electricity generation in Egypt heavily depends on natural gas and oil, which together account for over 90% of the country's energy consumption. However, the rapid growth in energy demand poses significant challenges to the stability of the national grid, risking potential supply–demand imbalances. To ensure long-term sustainability and conserve fossil fuel reserves, diversifying Egypt's energy mix by incorporating a higher proportion of renewable energy (RE) sources is essential.

Numerous studies have highlighted the economic and technical viability of hybrid energy systems, particularly those combining photovoltaic (PV) systems, wind turbines (WTs), fuel cells, electrolyzers, battery storage, and backup diesel generators (DGs). These hybrid systems aim to maximize RE utilization by tailoring system configurations to local resource availability and load requirements.

This study focuses on assessing the RE potential in a specific region of Egypt, with an emphasis on solar and wind resources, to evaluate the feasibility of sustainable energy production. Limited resources such as biomass, wave, and tidal energy are excluded from the analysis due to their scarcity in the region. The research explores various RE-based distributed generation systems, starting with the determination of optimal component sizes and costs, followed by the development of effective operational strategies.

The configurations under consideration include PV/battery and PV/WT/battery systems, with a conventional diesel-based system used for comparison. Diesel-based hybrid systems are deliberately excluded to prioritize the development of fully RE alternatives. A detailed comparative analysis is conducted between RE-based systems and traditional diesel generation, focusing on economic indicators such as net present cost (NPC), levelized cost of energy (LCOE), and energy performance metrics. By examining these systems, the study aims to identify the most cost-effective and sustainable energy solution for the selected region, contributing to Egypt's transition toward a more diversified and resilient energy mix.

6.2   Review

Numerous studies have been conducted to identify the most efficient RE sources for alternative power generation systems, considering factors such as resource availability and the specific energy demands of a location. PV systems, known for their renewable nature and minimal environmental impact, are frequently paired with battery storage and have gained global adoption [1,2]. However, PV generation depends on variables like solar irradiance, temperature, and daylight hours [3], requiring energy storage to compensate for the lack of sunlight during nighttime. This necessitates larger battery capacities to ensure a stable power supply. While DGs are commonly used as backup power sources for critical loads, they come with significant fuel costs and logistical challenges, particularly in remote areas where transporting fuel can be difficult.

To address these challenges, several energy sources, such as PV, wind, biomass, hydropower, and fuel cells, have been explored through the integration of hybrid renewable energy systems (HRES). While biomass and hydropower offer promising solutions, they require significant infrastructure and a reliable fuel supply, making them less practical in certain contexts. Conversely, simpler alternatives like PV and wind hybrid systems combined with battery storage present a more feasible solution. However, both PV and wind energy are susceptible to fluctuations in supply due to varying weather conditions, highlighting the need for reliable energy storage and system optimization to maintain a consistent power supply [4–6].

To reduce greenhouse gas emissions, a proposed HRES has been designed for a region in Russia with peak loads ranging from 10 to 100 kW. This system, which combines solar and wind energy with battery storage, has proven to be the most efficient solution for short-term power needs. The system includes PV modules, WTs, and infrastructure for hydrogen production and storage. However, when solar and wind resources were insufficient, the cost of hydrogen storage increased, highlighting the challenges associated with balancing intermittent renewable generation with storage costs [7,8].

A feasibility study for large-scale tourist operations identified the most economically viable RE source, which outperformed PV in such applications. Additionally, an advanced particle swarm optimization technique was used to optimize the capacity allocation in standalone hybrid wind/solar/battery systems. This approach aimed to balance environmental sustainability with economic efficiency by applying eco-design optimization to autonomous hybrid systems that integrate wind, PV, and battery storage. To account for energy source correlations in hybrid wind–PV systems, a two-point estimation method was employed, and a novel methodology for probabilistic power flow analysis was introduced. Furthermore, the optimal contract and installed capacities for wind and PV generation were analyzed, taking into consideration the time-of-use rates for industrial consumers [9–15].

Although many studies share similar objectives, only a limited number focus on designing and comparing configurations that integrate RE with battery storage. This study aims to evaluate two RE systems, incorporating either hydrogen or battery storage, to identify the most cost-effective design that maximizes RE potential. The analysis is conducted for a typical household in New Minya, Egypt, considering the region's diverse climatic conditions. Using HOMER (Hybrid Optimization Model for Electric Renewables), a state-of-the-art software tool, the study assesses the reliability and performance of hybrid systems that combine RE sources, such as PV and wind, with battery storage.

6.3 Hybrid energy system under study

The integration of RE sources into a multi-source hybrid system, configured as an off-grid microgrid, provides an efficient solution for optimizing electricity generation from both technical and economic perspectives. However, one of the key challenges in designing such systems is optimizing the size of energy components to balance cost-effective energy production with dependable system performance and reliability [16].

This study evaluates a hybrid energy system consisting of PV panels and WTs, supported by a battery storage system (BSS) to ensure uninterrupted power supply during times of low RE output. The system is specifically designed to address the electricity demands in Egypt, harnessing the region's RE potential to provide sustainable and reliable power.

6.3.1   Standalone hybrid renewable energy system

The goal of this study is to design an optimal standalone hybrid generation system from an economic standpoint, while adhering to operational constraints [16]. The total annual cost encompasses the initial investment, operation and maintenance (O&M) expenses for each power source, along with the salvage value of the equipment, which is deducted. The NPC serves as the primary evaluation criterion, and the optimal configuration, determined using HOMER software, has been thoroughly analyzed [17]. Moreover, Table 6.1 provides the component data for the three systems. The detailed specifications can be summarized [6,8,16,18,19].

This study incorporates several components to create an HRES. The PV system used is the Suntech Ltd. model, with PV panels arranged on a horizontal axis and adjusted monthly. The associated capital, replacement, and O&M costs are $1,000, $1,000, and $10 per kW/year, respectively. For wind power, the SW Whisper 500 model, with a rated capacity of 10 kW and a rotor diameter of 4.5 m, is selected, with costs of $900 for capital and replacement and $36 for O&M per kW/year. The BSS is modeled as a two-tank energy storage unit using the Polarium SLB48-200-146-2 model (50.8 V, 200 Ah, 10.2 kWh), with capital and replacement costs of $1150 and O&M costs of $2 annually. A converter, necessary for systems with DC components supplying AC loads (or vice versa), is included, with costs of $500 for capital and replacement and $1 for O&M per kW/year.

Table 6.1   Technical specifications and study assumptions

Description	Data	Description	Data
PV—Generic flat plate PV		Wind—Generic 3 kW	
Capital cost	1,000 US $/kW	Capital cost	900 US $/kW
Lifetime	25 years	Lifetime	25 years
O&M cost	10 US $/kW/year	O&M cost	36 US $/kW/year
Inverter		Batteries	
Capital	500 US $/kW	Type of batteries	Fortress Power LFP-10
Lifetime	15 years	Nominal voltage (V)	50.8 V
O&M cost	0 US $/year	Nominal capacity (kWh)	10.2
		Nominal capacity (Ah)	200
		O&M	2$/year
		Cost	1,150 $
		Lifetime	15 years

6.3.2 Configurations under study

Three distinct configurations of a 100% RE system have been thoroughly analyzed in this study. The first configuration consists of a PV system paired with a battery storage solution, designed to harness solar energy and store it for use during periods of low sunlight. The second configuration incorporates a combination of PV, WT, and battery storage, allowing for a more diverse energy generation approach by combining both solar and wind power with the added flexibility of energy storage. All three configurations have been modeled using HOMER, a widely used software for optimizing RE systems. The detailed schematics and layout of these three configurations can be seen in Figures 6.1 and 6.2, which provide visual representations of how each system is structured and integrated.

6.3.3 Site data

The site data for the location in New Minya, Egypt, is presented in Figures 6.3–6.5, which show the wind speed, solar radiation, and temperature profiles, respectively. The average annual Global Horizontal Irradiance (GHI) for solar radiation is 5.92 kWh/m^2/day, indicating a favorable solar energy potential for the region.

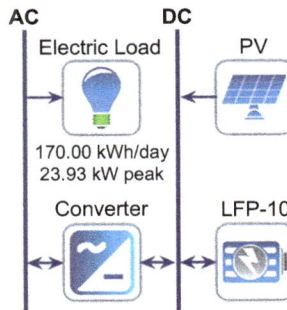

Figure 6.1 Configuration of the first system of PV/BSS

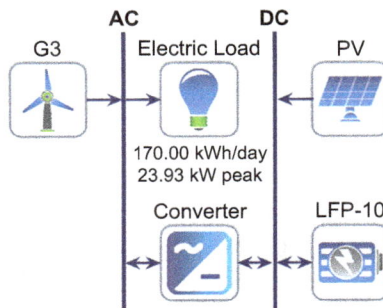

Figure 6.2 Configuration of the first system of PV/WT/BSS

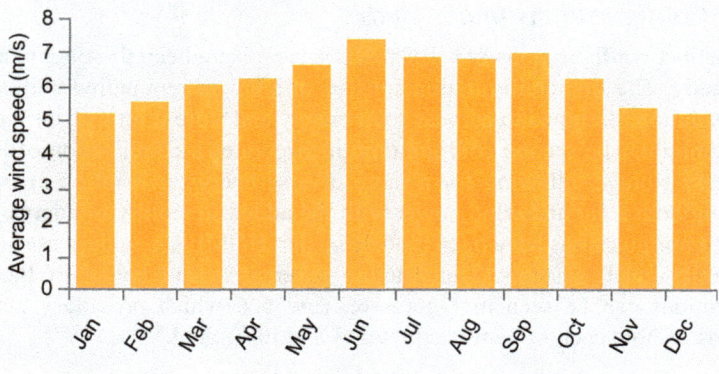

Figure 6.3   Average wind speed at the tested location

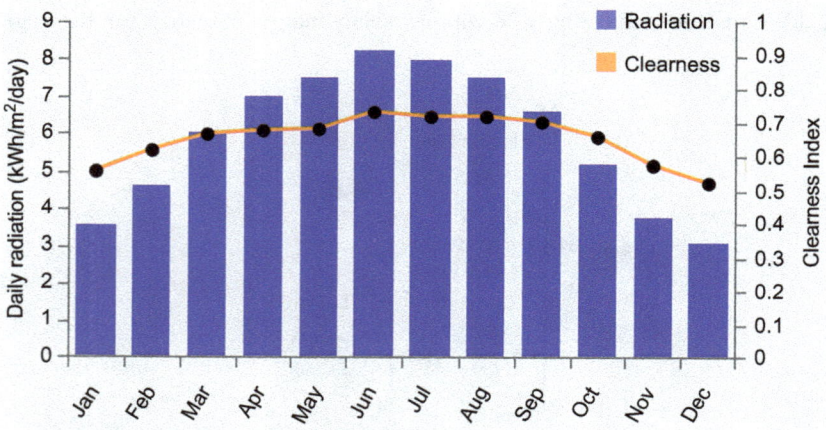

Figure 6.4   Daily radiation at the tested location

Additionally, the average annual wind speed is 6.2 m/s, suggesting a moderate wind resource that can complement solar generation. The annual average temperature is 22.58 °C, providing a temperate climate that could support efficient energy production and system operation.

Figure 6.6 illustrates the daily profile of the commercial load demand, while Figure 6.7 displays the seasonal load profile in kW. The peak load reaches 23.93 kW, with an average load of 7.08 kW. The total annual load is 170 kWh/day, indicating the energy needs of the commercial facility throughout the year. This data serves as the foundation for designing and optimizing the HRESs discussed in the study.

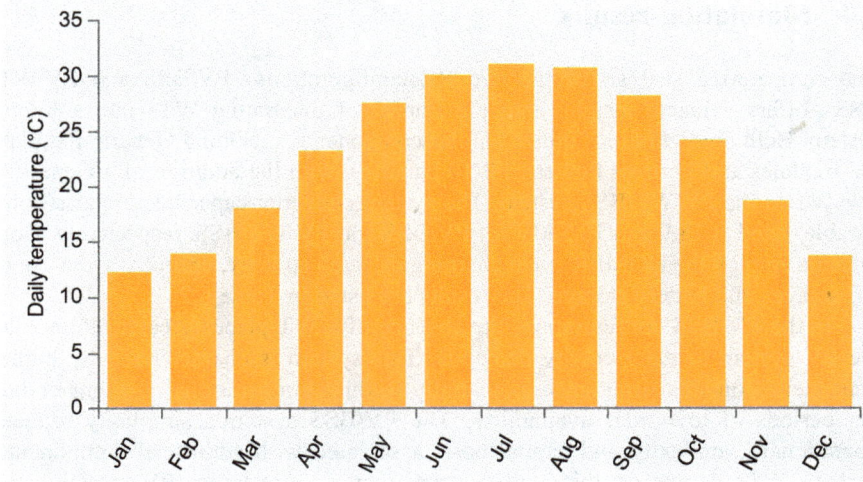

Figure 6.5   *The daily temperature at the tested location*

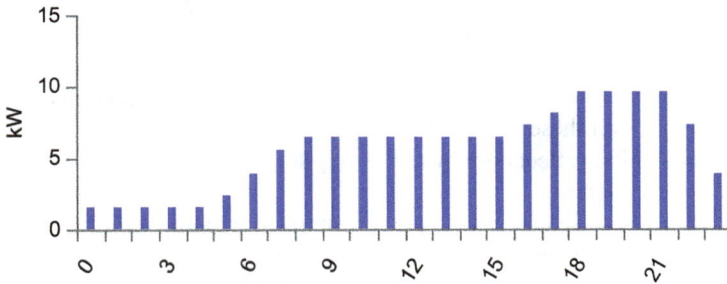

Figure 6.6   *Daily profile of the load demand*

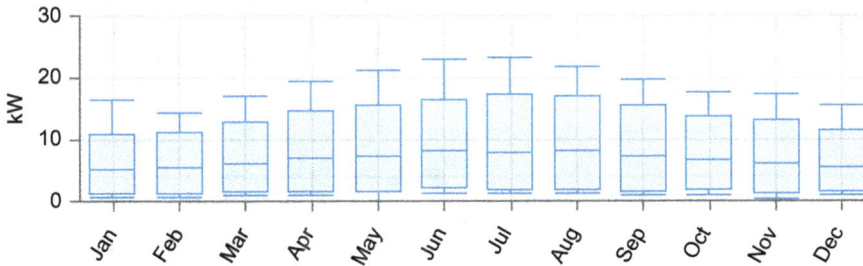

Figure 6.7   *Seasonal load profile in kW*

6.4   Simulation results

The comparative analysis of the two RE configurations—PV/BSS and PV/WT/
BSS—offers valuable insights into the impact of integrating WTs into a hybrid
system. Both configurations utilize similar components, including Generic flat plate
PV modules and Fortress Power LFP-10 batteries, with the addition of a Generic 3
kW WT in the PV/WT/BSS system. The system converter capacities are also comparable, rated at 18.9 and 18.5 kW for PV/BSS and PV/WT/BSS, respectively. Both
systems employ the HOMER cycle charging dispatch strategy, optimizing the use of
RE sources and battery storage. The results are listed in Table 6.2.

In the PV/BSS configuration, the absence of a WT places greater reliance on
the PV system to meet energy demands. This system is characterized by higher
dependency on solar irradiance, potentially leading to operational challenges during periods of low solar availability. The PV/BSS system's simplicity reduces
operational complexity and capital costs associated with additional components,
such as WTs. However, this configuration may require larger PV capacity and
battery storage to ensure reliability, particularly during seasonal or diurnal variations in solar energy availability.

The PV/WT/BSS configuration, conversely, benefits from the inclusion of the
WT, which provides an additional RE source. This diversification enhances the
system's ability to generate energy during periods of low solar irradiance, such as
at night or during overcast weather conditions. The integration of wind energy
reduces the dependency on PV panels and battery storage, potentially decreasing
operational stress on these components. The smaller converter capacity (18.5 kW)
in this configuration reflects optimized sizing due to the complementary nature of
wind and solar generation.

The HOMER cycle charging strategy employed in both configurations ensures
efficient energy management by prioritizing battery charging during surplus RE
production. This strategy minimizes reliance on backup systems and ensures optimal use of renewable resources. In the PV/WT/BSS configuration, the combined
use of solar and wind energy allows for more balanced charging and discharging
cycles in the BSS, improving its overall longevity and efficiency.

While the PV/WT/BSS configuration offers enhanced energy reliability and
reduced over-sizing of PV and storage components, it introduces additional capital
costs and maintenance requirements associated with the WT. The choice between

Table 6.2   Optimal systems configuration

Component	Name	PV/BSS	PV/WT/BSS
PV	Generic flat plate PV	44.9	44.6
Storage	Fortress Power LFP-10	15	15
WT	Generic 3 kW	–	1
System converter (kW)	System converter	18.9	18.5
Dispatch strategy	HOMER cycle charging		

these configurations should therefore consider site-specific factors, including wind resource availability, solar irradiance levels, and economic constraints. Ultimately, the integration of WTs in the PV/WT/BSS configuration provides a more resilient and versatile system, particularly in regions with complementary solar and wind resources.

The performance indices of the PV/BSS and PV/WT/BSS configurations reveal key economic and environmental aspects that highlight their feasibility and trade-offs in RE systems. Both configurations achieve a renewable fraction (RF) of 100%, indicating their complete reliance on RE sources and eliminating carbon dioxide (CO_2) emissions and fuel consumption. This underscores their environmental sustainability and suitability for off-grid or carbon-neutral energy applications. The results are listed in Table 6.3.

From an economic perspective, the NPC is slightly higher for the PV/WT/BSS configuration, amounting to $91,517 compared to $88,429 for the PV/BSS system. This increase reflects the additional capital expenditure associated with integrating the WT, as evidenced by the higher Capital Expenditure (CAPEX) of $70,113 in the PV/WT/BSS configuration versus $67,771 for PV/BSS. Similarly, the Operating Expenditure (OPEX) is marginally higher for PV/WT/BSS, at $1,674 compared to $1,616 for PV/BSS. These differences are attributed to the inclusion of the WT, which adds complexity and minor additional operational requirements.

In terms of cost-effectiveness, the LCOE is slightly higher for the PV/WT/BSS system at $0.119/kWh, compared to $0.116/kWh for PV/BSS. This increase is minimal and demonstrates that the integration of WTs does not drastically impact the overall economic efficiency of the system. The PV/BSS configuration thus maintains a slight advantage in terms of LCOE, primarily due to its simpler design and lower initial costs.

Both configurations achieve remarkable environmental performance, with zero fuel consumption and no CO_2 emissions, making them highly desirable for sustainable energy systems. The slight cost premium of the PV/WT/BSS system may be justified in scenarios where wind energy enhances system reliability or mitigates risks associated with solar variability.

Overall, while the PV/BSS configuration offers a more cost-effective solution for areas with high solar irradiance, the PV/WT/BSS configuration provides added

Table 6.3   Indices of the tested configurations

	PV/BSS	PV/WT/BSS
NPC	$88,429	$91,517
CAPEX	$67,771	$70,113
OPEX	$1,616	$1,674
LCOE (per kWh)	$0.116	$0.119
RF %	100	100
CO_2 emitted (kg/year)	0	0
Fuel consumption (L/year)	0	0

resilience through resource diversification. The choice between these systems depends on site-specific resource availability, economic considerations, and system reliability requirements.

The energy results for the PV/BSS and PV/WT/BSS configurations highlight their operational performance, including energy generation, utilization, and efficiency. The results are listed in Table 6.4.

The total energy generation is nearly identical for both configurations, with the PV/BSS system producing 82,229 kWh/year and the PV/WT/BSS system producing a slightly higher 82,446 kWh/year. This marginal increase reflects the contribution of the WT, which generates 650 kWh/year in the PV/WT/BSS configuration. However, the PV generation is slightly lower in the PV/WT/BSS system (81,796 kWh/year) compared to PV/BSS (82,229 kWh/year). This reduction could stem from variations in system optimization or reduced reliance on PV as the WT supplements energy needs.

The excess electricity produced is comparable between the two configurations, with 18,685 kWh/year for PV/BSS and 18,897 kWh/year for PV/WT/BSS. This indicates that both systems have a similar degree of overproduction, which could be leveraged for future expansion or auxiliary uses, such as energy storage or hydrogen production.

The unmet electric load is slightly lower in the PV/WT/BSS system (2,122 kWh/year) compared to PV/BSS (2,163 kWh/year). Similarly, the capacity shortage follows the same trend, being marginally lower in the PV/WT/BSS system (3,123 kWh/year) compared to PV/BSS (3,147 kWh/year). These results suggest that integrating a WT can enhance the system's ability to meet demand, albeit with minor improvements.

The AC primary load is effectively met in both configurations, with 59,887 kWh/year for PV/BSS and 59,928 kWh/year for PV/WT/BSS. The minimal difference between these values demonstrates that both systems are well-designed to satisfy the load requirements.

Overall, the addition of a WT in the PV/WT/BSS configuration slightly improves system reliability by reducing unmet load and capacity shortages. However, the benefits are relatively small, which may not justify the additional cost

Table 6.4   Energy results of the analyzed three configurations

Quantity	PV/BSS	PV/WT/BSS
Excess electricity (kWh/year)	18,685	18,897
Unmet electric load (kWh/year)	2,163	2,122
Capacity shortage (kWh/year)	3,147	3,123
Generic flat plate PV (kWh/year)	82,229	81,796
Generic 3 kW (kWh/year)	–	650
Total (kWh/year)	82,229	82,446
AC primary load (kWh/year)	59,887	59,928
Total	59,887	59,928

and complexity unless wind resources are particularly favorable at the installation site. The choice of configuration should therefore consider the trade-offs between enhanced reliability and the economic implications of incorporating wind energy.

6.4.1   Configuration 1: PV/BSS system

The results of the PV/BSS system analysis highlight the economic and technical performance of the configuration, providing key insights into its feasibility and areas for improvement. The results are listed in Table 6.5. The economic break-down reveals a net present cost (NPC) of $88,429, with the majority of costs stemming from the capital investment in the PV system ($44,861) and the BSS ($17,250). Replacement costs, primarily driven by the BSS, further contribute to lifecycle expenses, underscoring the importance of extending battery life to reduce long-term costs. The PV system's cost-effectiveness, with an LCOE of $0.0481/ kWh, aligns well with RE benchmarks, ensuring competitive energy generation over its operational life. Figure 6.8 provides a detailed breakdown of the net present value (NPV) for each component in the PV/BSS system. Panel (a) illustrates the total NPV of the system, capturing the overall economic viability and cost-effectiveness of the configuration over its lifetime. Panel (b) delves deeper into the costs associated with individual components, such as the PV panels, BSS, and system converter. Figure 6.9 shows the average monthly electric energy generation from renewable sources in the first configuration PV/BSS.

From a technical perspective, the PV system's performance is robust, produc-ing 82,229 kWh/year at a capacity factor of 20.9%, as listed in Table 6.6. This output exceeds load requirements, with a penetration rate of 133%, indicating potential over-sizing. This surplus energy could be utilized for additional applica-tions, such as load expansion, increased storage capacity, or hydrogen production. The system's efficiency is further demonstrated by the absence of clipped pro-duction, maximizing energy harvesting from the PV panels. However, the over-sizing also results in the underutilization of other components, such as the battery and converter, highlighting the need for optimization in system design.

The BSS exhibits reasonable performance, with 137 kWh of usable capacity and an autonomy of 19.3 h, ensuring reliability during periods of low generation as shown in Table 6.7. The annual equivalent full cycles (EFCs) of 221 cycles/year reflect improved utilization compared to typical configurations, suggesting a better match with the system's energy demands. Despite this, energy losses of 642 kWh/

Table 6.5   NPC of the PV/BSS system

Name	Capital	Operating	Replacement	Salvage	Total
Fortress Power LFP-10	$17,250	$0.00	$15,011	−$2,010	$30,251
Generic flat plate PV	$44,861	$5,735	$0.00	$0.00	$50,596
System converter	$5,659	$0.00	$2,361	−$439.55	$7,581
System	$67,771	$5,735	$17,372	−$2,449	$88,429

(a)

NPC

($20,000)

■ Generic flat plate PV   ■ Polarium SLB48-200-146-2   ■ System Converter

(b)

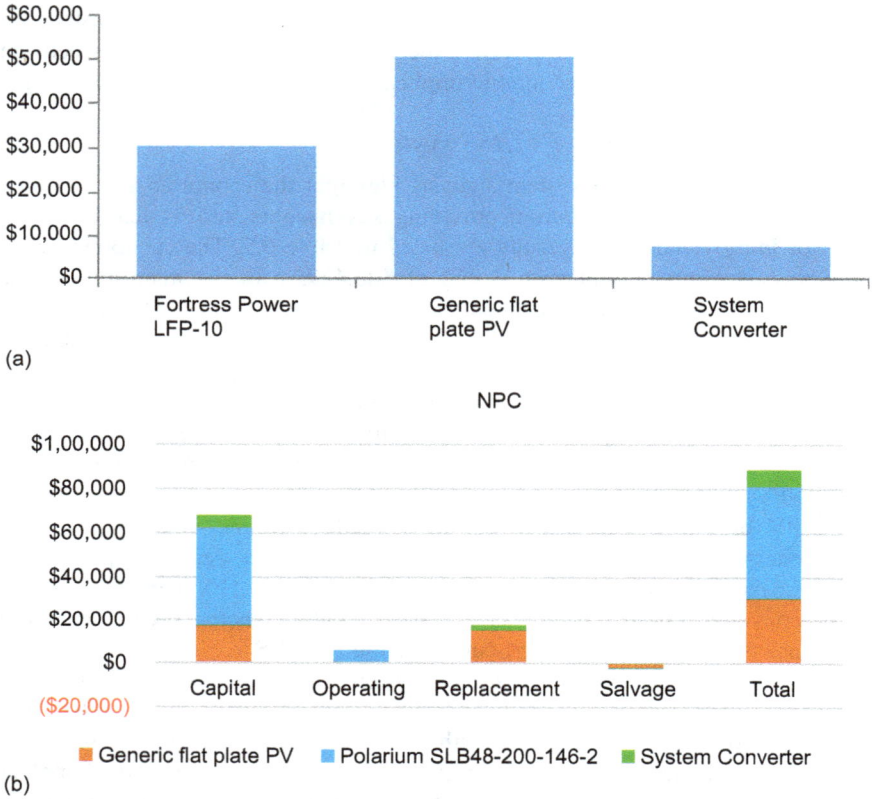

Figure 6.8   NPV for each system component for the PV/BSS system: (a) total NPV and (b) components costs

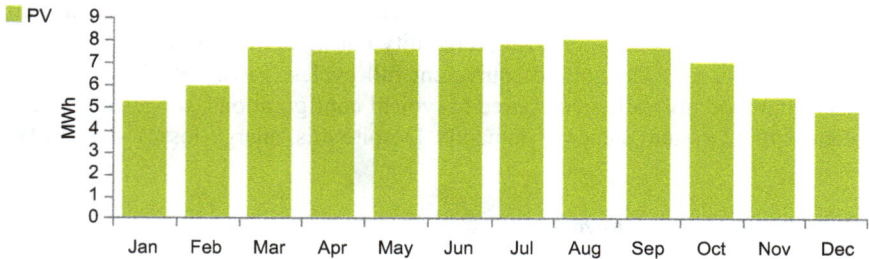

Figure 6.9   Average monthly electric energy generation from renewable sources in the first configuration PV/BSS

Table 6.6   Performance and indices of PV for the PV/BSS system

Quantity	Value	Unit
Rated capacity	44.9	kW
Mean output	9.39	kW
Mean output	225	kWh/day
Capacity factor	20.9	%
Total production	82,229	kWh/year
Minimum output	0	kW
Maximum output	44.1	kW
PV penetration	133	%
Hours of operation	4,382	h/year
Levelized cost	0.0481	$/kWh
Clipped production	0	kWh

Table 6.7   Performance and indices of BSS system for the PV/BSS system

Quantity	Value	Unit
Batteries	15.0	qty.
String size	1.00	batteries
Strings in parallel	15.0	strings
Bus voltage	48.0	V
Autonomy	19.3	h
Storage wear cost	0.0290	$/kWh
Nominal capacity	144	kWh
Usable nominal capacity	137	kWh
Lifetime throughput	318,670	kWh
Expected life	10.0	kWh
Energy in	32,052	kWh/year
Energy out	31,547	kWh/year
Storage depletion	137	kWh/year
Losses	642	kWh/year
Annual throughput	31,867	kWh/year
Annual EFCs	221	1/year
Average daily EFCs	0.606	1/day

year (~2%) and storage depletion highlight areas where efficiency could be improved. Additionally, the low storage wear cost of $0.029/kWh underscores the cost-effectiveness of the BSS over its lifetime, but its finite lifespan necessitates future replacements, adding to the system's overall costs. Figure 6.10 shows the average monthly state of charge for the first configuration PV/BSS.

The converter, rated at 18.9 kW, achieves a capacity factor of 36.2% and operates for 8,385 hours/year, indicating effective use. However, energy losses of 3,152 kWh/year (~5%) suggest room for improvement, potentially through the adoption of advanced power conversion technologies as listed in Table 6.8. The

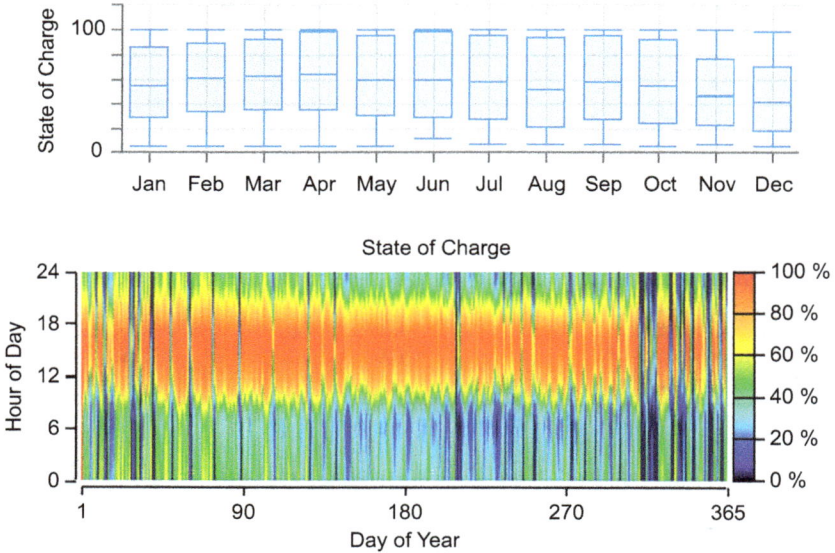

Figure 6.10   Average monthly state of charge for the first configuration PV/BSS

*Table 6.8   Performance and indices of converter for the
PV/BSS system*

Quantity	Inverter	Rectifier	Unit
Capacity	18.9	18.9	kW
Mean output	6.84	0	kW
Minimum output	0	0	kW
Maximum output	18.9	0	kW
Capacity factor	36.2	0	%
Hours of operation	8,385	0	h/year
Energy out	59,887	0	kWh/year
Energy in	63,039	0	kWh/year
Losses	3,152	0	kWh/year

converter's high reliability and consistent operation play a critical role in maintaining energy flow, but its high rating relative to demand may contribute to inefficiencies, suggesting possible over-sizing.

Overall, the system demonstrates strong economic viability and technical efficiency, with clear areas for refinement. The PV system is cost-effective and provides surplus energy, while the BSS and converter offer reliable performance with manageable inefficiencies. However, optimizing component sizes, particularly the battery and converter, could reduce underutilization and further enhance the system's overall performance. These adjustments could unlock additional economic and environmental benefits, making the PV/BSS system even more sustainable and cost-effective in the long term.

The performance of the PV/BSS configuration is detailed in Figure 6.11, which presents the charging and discharging patterns of the BSS over a two-day period. The figure demonstrates how the BSS effectively stores excess energy generated by the PV panels during peak production hours and discharges it to meet the load demand during periods of lower generation. Additionally, Figure 6.12 illustrates the energy balance for a single day, showcasing the interaction between the PV panels and the BSS in maintaining a stable energy supply. These figures highlight the configuration's ability to manage energy storage and ensure reliable power delivery in the absence of additional renewable sources.

6.4.2   Configuration 2: PV//WT/BSS system

The PV/WT/BSS system demonstrates its capability to meet load demand effectively while achieving notable cost and operational performance improvements. Table 6.9 presents the NPC breakdown for the system, amounting to $91,517. The primary contributors to capital costs include the PV system ($44,625) and the BSS ($17,250), together forming the majority of the initial investment. Operating expenses are minimal at $6,152, while replacement costs amount to $18,164, indicating moderate ongoing expenditures. The salvage value of −$2,912 slightly offsets the total cost, reflecting the potential for end-of-life system recovery. The results underscore the system's affordability and its potential for cost optimization through enhanced component longevity and operational efficiency. These results are visualized in Figure 6.13.

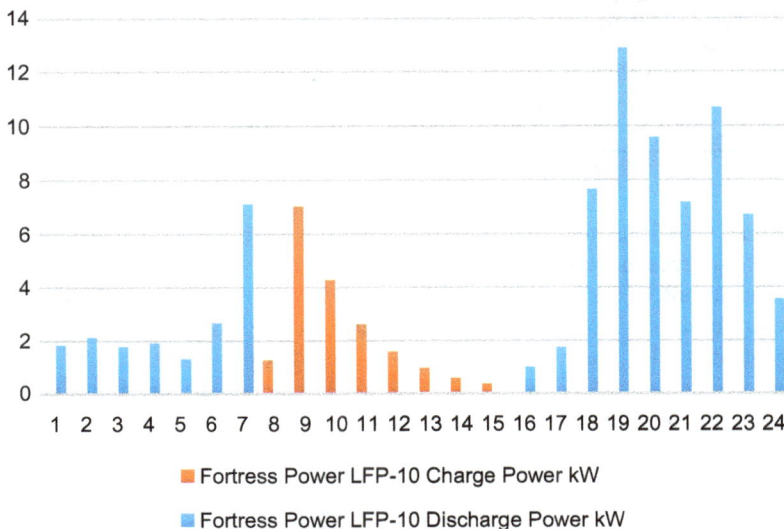

Figure 6.11   BSS charging and discharging for two days for the first configuration PV/BSS

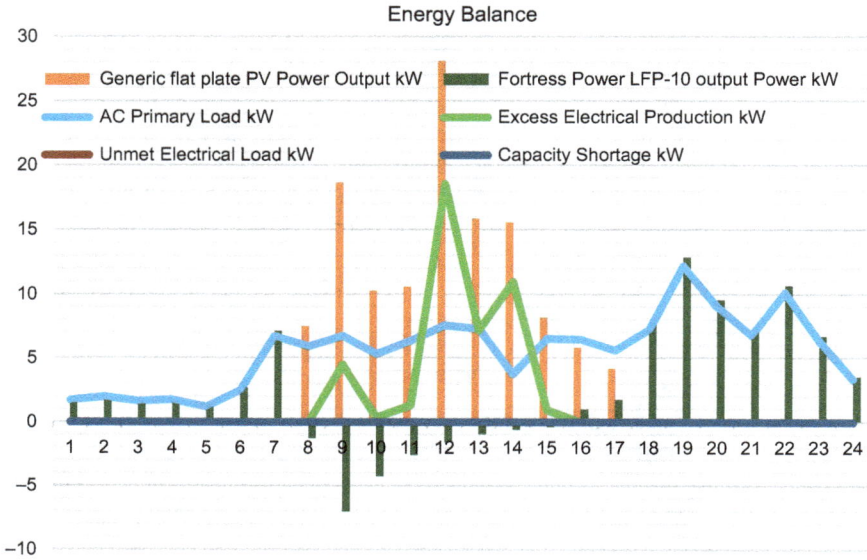

Figure 6.12   The energy balance of the first configuration for one day PV/BSS

Table 6.9   NPCs for PV//WT/BSS system

Name	Capital	Operating	Replacement	Salvage	Total
Fortress Power LFP-10	$17,250	$0.00	$15,011	−$2,010	$30,251
Generic 3 kW	$2,700	$447.42	$841.87	−$471.82	$3,517
Generic flat plate PV	$44,625	$5,705	$0.00	$0.00	$50,330
System converter	$5,538	$0.00	$2,311	−$430.11	$7,419
System	$70,113	$6,152	$18,164	−$2,912	$91,517

Figure 6.14 illustrates the average monthly electric energy generation from renewable sources in the second configuration (PV/WT/BSS). The figure highlights the contributions of PV panels and WTs to the overall energy production across different months. It showcases seasonal variations, with higher energy generation during months with favorable solar irradiance and wind conditions, underscoring the complementary nature of the two renewable sources in ensuring a consistent energy supply throughout the year. This analysis demonstrates the capability of the PV/WT/BSS configuration to adapt to varying resource availability and maintain energy reliability.

Battery storage system (BSS) performance metrics, detailed in Table 6.10, highlight the system's reliability and energy management capabilities. The BSS provides an autonomy of 19.3 h with a nominal capacity of 144 kWh and a usable capacity of 137 kWh. The storage wear cost is competitive at $0.0290/kWh, with a

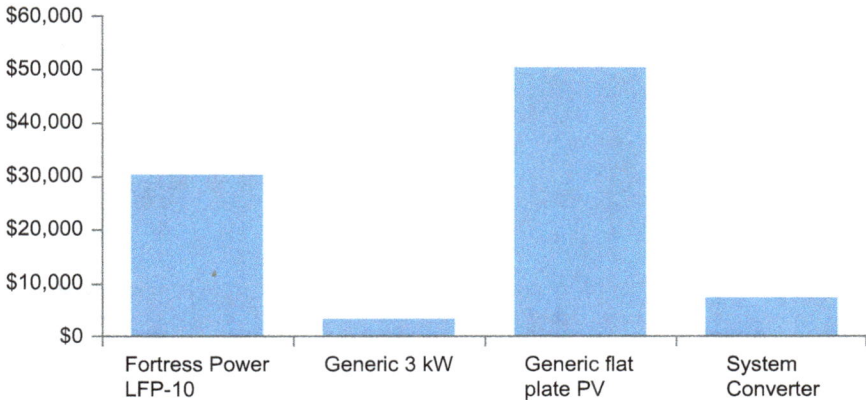

(a)

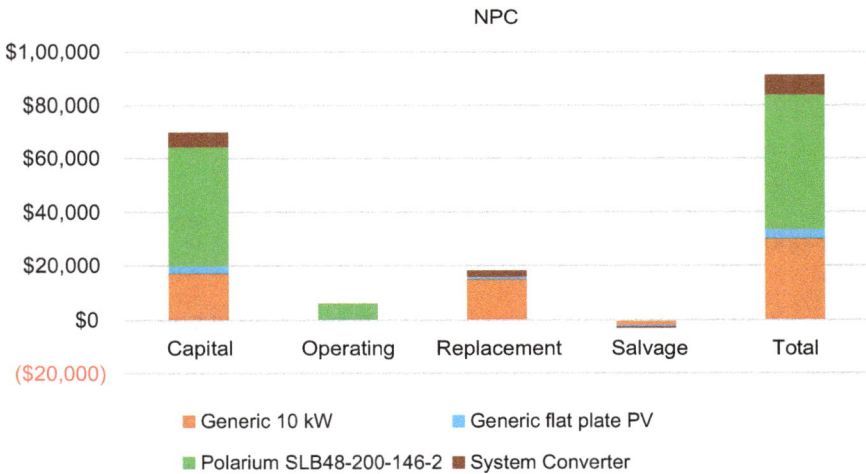

(b)

Figure 6.13   *NPV for each system component for the PV/WT/BSS system: (a) total NPV and (b) components costs*

lifetime throughput of 316,292 kWh over an expected life of 10 years. Annual energy losses are minimal at 137 kWh, and the BSS supports 31,629 EFCs per year, averaging 220 EFCs daily. These metrics confirm the BSS's efficiency in balancing RE intermittency, although minor losses indicate scope for further optimization. The results of the state of charge are presented in Figure 6.15.

Table 6.11 outlines the PV system's performance, with a rated capacity of 44.6 kW and a capacity factor of 20.9%. The system generates an annual production of 81,796 kWh at a levelized cost of $0.0481/kWh. It operates for 4,382 h annually, with no clipped production, ensuring efficient utilization of installed capacity. PV

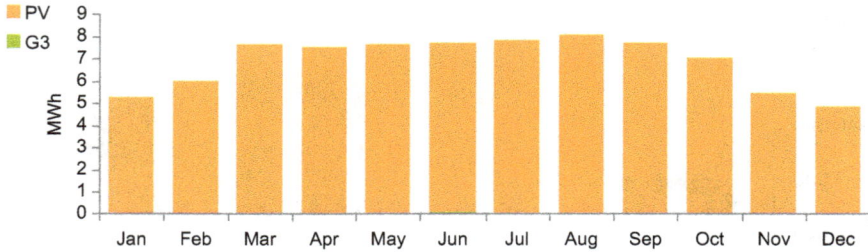

Figure 6.14   Average monthly electric energy generation from renewable sources in the second configuration PV/WT/BSS

Table 6.10   BSS performance for PV//WT/BSS system

Quantity	Value	Unit
Batteries	15.0	qty.
String size	1.00	batteries
Strings in parallel	15.0	strings
Bus voltage	48.0	V
Autonomy	19.3	h
Storage wear cost	0.0290	$/kWh
Nominal capacity	144	kWh
Usable nominal capacity	137	kWh
Lifetime throughput	316,292	kWh
Expected life	10.0	year
Energy in	0	kWh/year
Energy out	31,812	kWh/year
Storage depletion	31,311	kWh/year
Losses	137	kWh/year
Annual throughput	638	kWh/year
Annual EFCs	31,629	1/year
Average daily EFCs	220	1/day

penetration of 132% highlights its substantial contribution to meeting load requirements, making it a key component of the hybrid configuration.

The WT system, detailed in Table 6.12, complements the PV system with a total rated capacity of 3 kW and an annual production of 650 kWh. Despite its modest capacity factor of 2.47%, the WT system contributes to overall reliability by generating energy during periods of low solar irradiance. However, its wind penetration of 1.05% and higher levelized cost of $0.424/kWh suggest limited economic viability. These results emphasize the WT system's secondary role in the configuration, primarily enhancing resilience rather than providing significant cost savings.

The converter's performance, as shown in Table 6.13, emphasizes its critical role in integrating and managing energy flows from the hybrid system. The inverter and rectifier are each rated at 18.5 kW, with the inverter achieving a capacity factor

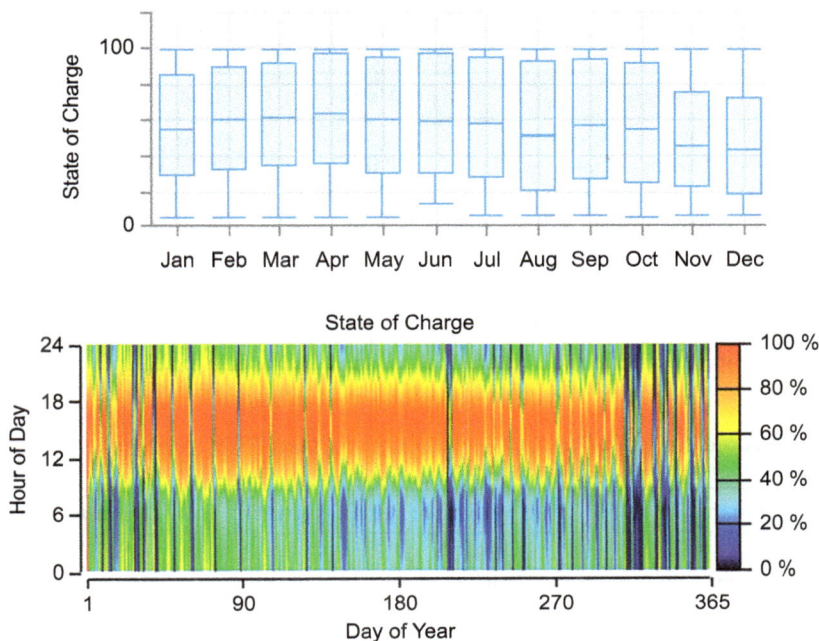

Figure 6.15 Average monthly state of charge for the second configuration PV/WT/BSS

Table 6.11 Performance of PV units for PV//WT/BSS system

Quantity	Value	Unit
Rated capacity	44.6	kW
Mean output	9.34	kW
Mean output	224	kWh/day
Capacity factor	20.9	%
Total production	81,796	kWh/year
Minimum output	0	kW
Maximum output	43.9	kW
PV penetration	132	%
Hours of operation	4,382	h/year
Levelized cost	0.0481	$/kWh
Clipped production	0	kWh

of 36.7%. Operating for 8,400 h annually, the inverter outputs 59,278 kWh/year, with losses totaling 3,120 kWh/year. These losses, though minor, present opportunities for further efficiency improvements. The converter's performance ensures seamless integration of multiple energy sources, contributing to the overall system stability and energy continuity.

Table 6.12   Indices WT for the PV//WT/BSS system

Quantity	Value	Unit
Total rated capacity	3.00	kW
Mean output	0.0742	kW
Capacity factor	2.47	%
Total production	650	kWh/year
Minimum output	0	kW
Maximum output	2.55	kW
Wind penetration	1.05	%
Hours of operation	4,433	h/year
Levelized cost	0.424	$/kWh

Table 6.13   Indices of the converter for the PV//WT/BSS system

Quantity	Inverter	Rectifier	Quantity
Capacity	18.5	18.5	kW
Mean output	6.77	0	kW
Minimum output	0	0	kW
Maximum output	18.5	0	kW
Capacity factor	36.7	0	%
Hours of operation	8,400	0	h/year
Energy out	59,278	0	kWh/year
Energy in	62,398	0	kWh/year
Losses	3,120	0	kWh/year

Operational dynamics of the PV/WT/BSS system highlight its ability to effectively balance energy supply and demand. The BSS manages excess energy generated by PV panels and WTs during surplus periods, storing it for use during periods of insufficient generation. Figures illustrating energy flows and state-of-charge dynamics confirm the system's capability to maintain reliability and minimize unmet load. These results collectively underscore the cost-effectiveness, sustainability, and reliability of the PV/WT/BSS system, while also identifying areas for potential improvements in WT utilization and converter efficiency.

The operational dynamics of the PV/WT/BSS configuration are depicted in Figure 6.16, which illustrates the charging and discharging cycles of the BSS over a two-day period. The figure highlights how the BSS effectively manages energy by storing excess electricity generated by the PV panels and WTs during periods of surplus and discharging it during times of insufficient generation to meet the load demand. Complementing this, Figure 6.17 provides a detailed view of the energy balance for a single day, showcasing the interaction between the PV panels, WTs, and BSS in maintaining a stable and reliable energy supply. These figures collectively emphasize the configuration's ability to optimize RE utilization while minimizing unmet load.

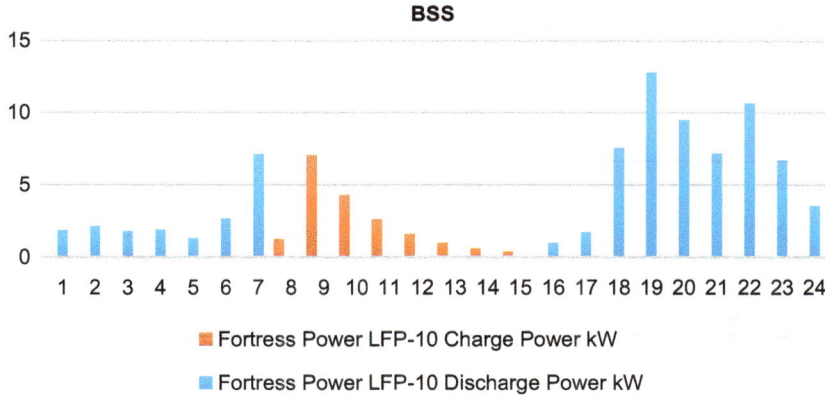

Figure 6.16   *BSS charging and discharging for two days for the second configuration PV/WT/BSS*

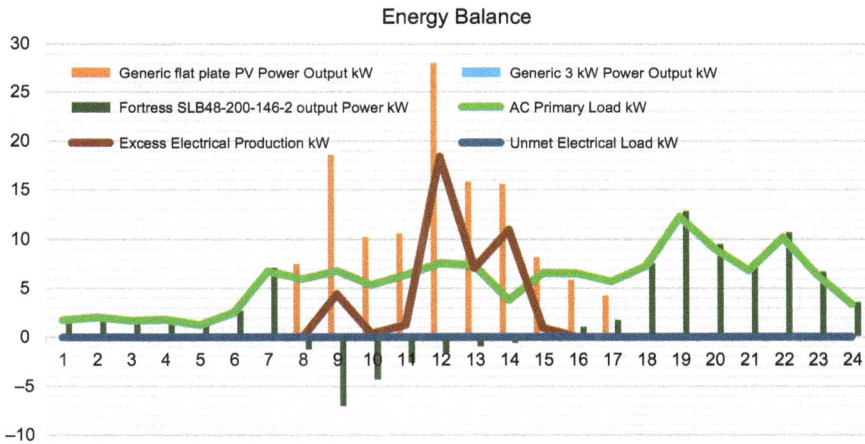

Figure 6.17   *The energy balance of the second configuration for one day PV/WT/ BSS*

6.4.3   *Comprehensive comparison with base case of DG*

The results of Table 6.14 demonstrate the clear economic and environmental advantages of transitioning from a DG-based system to RE systems, specifically the PV/BSS and PV/WT/BSS configurations. Both proposed systems significantly reduce the NPC compared to the DG system, with savings exceeding 76%. This reduction is largely attributed to the elimination of fuel costs and the reduced operational OPEX associated with RE systems. Among the renewable options, the PV/BSS system exhibits a slightly lower NPC, indicating superior cost-effectiveness over the PV/WT/BSS system.

Table 6.14   Comparison against the base case of DG only

	Base system of DG	Proposed system of PV/BSS	Proposed system of PV/WT/BSS
NPC	$396,922	$88,429	$91,517
CAPEX	$13,500	$67,771	$70,113
OPEX	$29,994	$1,616	$1,674
LCOE (per kWh)	$0.500	$0.116	$0.119
CO2 emitted (kg/year)	80,735	0	0
Fuel consumption (L/year)	30,843	0	0

From a financial perspective, the PV/BSS system outperforms the PV/WT/BSS system in terms of return on investment. It achieves a higher Internal Rate of Return (50.8% vs. 48.7%) and shorter payback periods, both discounted (2.09 years) and simple (1.91 years). These indicators highlight the PV/BSS system's capacity to provide quicker returns to investors while maintaining low operational costs. The PV/WT/BSS system, while slightly less competitive in financial metrics, offers additional benefits, such as improved energy reliability, particularly in regions with favorable wind conditions. This makes it a viable alternative for locations where wind can complement solar energy generation, potentially enhancing overall system performance and resilience.

The environmental benefits of both renewable systems are undeniable, with both eliminating CO_2 emissions and fuel consumption entirely. This is in stark contrast to the DG system, which emits 80,735 kg of CO_2 annually and consumes 30,843 L of fuel per year. These reductions in greenhouse gas emissions and fossil fuel dependency underscore the critical role of RE systems in promoting sustainability and addressing climate change.

The LCOE further highlights the economic efficiency of the renewable systems. Both PV/BSS and PV/WT/BSS achieve LCOEs of $0.116/kWh and $0.119/kWh, respectively, representing a reduction of over 76% compared to the DG system ($0.500/kWh). This reduction demonstrates the long-term affordability of RE systems, especially in areas where diesel fuel prices are high or volatile.

Despite the higher capital expenditure (CAPEX) of the renewable systems compared to the DG system, the substantial long-term cost savings and environmental benefits justify the initial investment. The PV/BSS system's lower CAPEX ($67,771) compared to the PV/WT/BSS system ($70,113) reflects its simpler design, as it does not require WT components. However, the inclusion of WTs in the PV/WT/BSS system may provide additional advantages in energy production diversification, potentially improving system reliability in varying weather conditions.

In conclusion, both the PV/BSS and PV/WT/BSS systems offer significant improvements over the DG system in terms of economic and environmental performance. The PV/BSS system emerges as the more cost-effective option, while the

PV/WT/BSS system provides additional reliability benefits in suitable locations. The results highlight the potential for RE systems to deliver sustainable, cost-effective, and reliable energy solutions, particularly in areas where diesel generation is currently the primary source of electricity.

6.5   Conclusions

This paper presents a comprehensive analysis of three RE configurations: the base case of a DG system, and two proposed systems comprising photovoltaic (PV) and battery storage (BSS), and PV with WT and battery storage (BSS). The analysis was performed using HOMER software to identify the optimal system configurations with the lowest net present cost (NPC) and LCOE, tailored for a commercial application in New Minya, Egypt. The comparison revealed that both the PV/BSS and PV/WT/BSS systems offer substantial economic advantages over the DG-only system. Specifically, the PV/BSS system has an NPC of \$88,429, a significant reduction from the DG-only system's NPC of \$396,922. The PV/WT/BSS system also showed a favorable NPC of \$91,517. Additionally, the capital expenditure (CAPEX) for the PV/BSS and PV/WT/BSS systems is considerably lower than that of the DG system, which has a CAPEX of \$13,500. The operating expenses (OPEX) for the renewable systems are much lower as well, with PV/BSS and PV/WT/BSS systems costing only \$1,616 and \$1,674 per year, respectively, compared to \$29,994 for the DG-only system. In terms of LCOE, the PV/BSS system provides a highly competitive rate of \$0.116/kWh, and the PV/WT/BSS system is slightly higher at \$0.119/kWh. These values are significantly lower than the DG-only system's LCOE of \$0.500/kWh, reinforcing the cost-effectiveness of the proposed RE configurations. Furthermore, both RE systems completely eliminate carbon emissions and fuel consumption, contributing zero CO_2 emissions and fuel consumption annually, a sharp contrast to the DG-only system, which emits 80,735 kg of CO_2 and consumes 30,843 L of fuel per year. Regarding energy generation, the PV/BSS system produces 82,229 kWh/year, while the PV/WT/BSS system produces slightly more at 82,446 kWh/year. These systems generate significant excess electricity (18,685 kWh/year for PV/BSS and 18,897 kWh/year for PV/WT/BSS), with minimal unmet load or capacity shortage. This overproduction can be used for storage or future scalability, including potential applications like hydrogen production. The PV/BSS and PV/WT/BSS configurations offer substantial benefits in terms of cost reduction, energy efficiency, and environmental impact when compared to the conventional DG-only system. The PV/BSS system stands out for its cost-effectiveness, while the PV/WT/BSS system provides slightly improved reliability in meeting energy demand. Further optimization of component sizes, as well as exploration of hydrogen storage and fuel cell integration, could enhance the economic feasibility and performance of these renewable systems, making them increasingly viable for large-scale commercial applications in regions like New Minya, Egypt. The transition to 100% HRESs will be fueled by advancements in storage technologies, artificial intelligence-driven optimization, and integration with other sectors such as hydrogen and electric vehicles. These systems will provide a more flexible, cost-effective, and sustainable energy infrastructure, enabling the world to meet growing energy demands while reducing reliance on fossil fuels. By harnessing the synergies between solar, wind, and storage technologies, hybrid systems will play a pivotal role in achieving global energy and climate goals.

References

[1]   M. Kolhe, S. Kolhe, and J. C. Joshi, "Economic viability of stand-alone solar photovoltaic system in comparison with diesel-powered system for India," *Energy Economics*, vol. 24, no. 2, pp. 155–165, 2002, doi:10.1016/s0140-9883(01)00095-0.

[2]   K. Y. Lau, C. W. Tan, and A. H. M. Yatim, "Photovoltaic systems for Malaysian islands: Effects of interest rates, diesel prices and load sizes," *Energy*, vol. 83, pp. 204–216, 2015, doi:10.1016/j.energy.2015.02.015.

[3]   E. Skoplaki and J. A. Palyvos, "On the temperature dependence of photovoltaic module electrical performance: A review of efficiency/power correlations," *Solar Energy*, vol. 83, no. 5, pp. 614–624, 2009, doi:10.1016/j.solener.2008.10.008.

[4]   B. Wichert, "PV-diesel hybrid energy systems for remote area power generation—A review of current practice and future developments," *Renewable and Sustainable Energy Reviews*, vol. 1, no. 3, pp. 209–228, 1997, doi:10.1016/s1364-0321(97)00006-3.

[5]   O. C. Onar, M. Uzunoglu, and M. S. Alam, "Dynamic modeling, design and simulation of a wind/fuel cell/ultra-capacitor-based hybrid power generation system," *Journal of Power Sources*, vol. 161, no. 1, pp. 707–722, 2006, doi:10.1016/j.jpowsour.2006.03.055.

[6]   H. S. Das, C. W. Tan, A. H. M. Yatim, and K. Y. Lau, "Feasibility analysis of hybrid photovoltaic/battery/fuel cell energy system for an indigenous residence in East Malaysia," *Renewable and Sustainable Energy Reviews*, vol. 76, pp. 1332–1347, 2017, doi:10.1016/j.rser.2017.01.174.

[7]   O. V. Marchenko and S. V. Solomin, "Modeling of hydrogen and electrical energy storages in wind/PV energy system on the Lake Baikal coast," *International Journal of Hydrogen Energy*, vol. 42, no. 15, pp. 9361–9370, 2017, doi:10.1016/j.ijhydene.2017.02.076.

[8]   M. Dekkiche, T. Tahri, and M. Denai, "Techno-economic comparative study of grid-connected PV/reformer/FC hybrid systems with distinct solar tracking systems," *Energy Conversion and Management: X*, vol. 18, p. 100360, 2023, doi:10.1016/j.ecmx.2023.100360.

[9]   A. A. Z. Diab, A. M. El-Rifaie, M. M. Zaky, and M. A. Tolba, "Optimal sizing of stand-alone microgrids based on recent metaheuristic algorithms," *Mathematics*, vol. 10, no. 1, p. 140, 2022, doi:10.3390/math10010140.

[10]   A. A. Z. Diab, H. M. Sultan, and O. N. Kuznetsov, "Optimal sizing of hybrid solar/wind/hydroelectric pumped storage energy system in Egypt based on

different meta-heuristic techniques," *Environmental Science and Pollution Research*, vol. 27, no. 26, pp. 32318–32340, 2019, doi:10.1007/s11356-019-06566-0.

[11] A. A. Z. Diab, H. M. Sultan, I. S. Mohamed, O. N. Kuznetsov, and T. D. Do, "Application of different optimization algorithms for optimal sizing of PV/wind/diesel/battery storage stand-alone hybrid microgrid," *IEEE Access*, vol. 7, pp. 119223–119245, 2019, doi:10.1109/access.2019.2936656.

[12] H. A. El-Sattar, S. Kamel, H. M. Sultan, H. M. Zawbaa, and F. Jurado, "Optimal design of photovoltaic, biomass, fuel cell, hydrogen tank units and electrolyzer hybrid system for a remote area in Egypt," *Energy Reports*, vol. 8, pp. 9506–9527, 2022, doi:10.1016/j.egyr.2022.07.060.

[13] A. M. Eltamaly, "A novel energy storage and demand side management for entire green smart grid system for <scp>NEOM</scp> city in Saudi Arabia," *Energy Storage*, vol. 6, no. 1, 2023, doi:10.1002/est2.515.

[14] A. M. Eltamaly, "Smart decentralized electric vehicle aggregators for optimal dispatch technologies," *Energies*, vol. 16, no. 24, p. 8112, 2023, doi:10.3390/en16248112.

[15] A. M. Eltamaly, "A novel energy storage and demand side management for entire green smart grid system for NEOM city in Saudi Arabia," *Energy Storage*, vol. 6, no. 1, p. e515, 2024.

[16] A. Al-Badi, A. Al Wahaibi, R. Ahshan, and A. Malik, "Techno-economic feasibility of a solar-wind-fuel cell energy system in Duqm, Oman," *Energies*, vol. 15, no. 15, p. 5379, 2022, doi:10.3390/en15155379.

[17] HOMER. "Hybrid Renewable and Distributed Generation System Design Software." Available from https://www.homerenergy.com/ [Accessed 7/2/2025].

[18] L. Abdolmaleki and U. Berardi, "Comparative analysis among three solar energy-based systems with hydrogen and electrical battery storage in single houses," Applied Energy Innovation Institute (AEii), 2022.

[19] V. Khare, S. Nema, and P. Baredar, "Optimization of hydrogen based hybrid renewable energy system using HOMER, BB-BC and GAMBIT," *International Journal of Hydrogen Energy*, vol. 41, no. 38, pp. 16743–16751, 2016, doi:10.1016/j.ijhydene.2016.06.228.

Index

www.ingramcontent.com/pod-product-compliance
Lightning Source LLC
Chambersburg PA
CBHW050514190326
41458CB00005B/1538